Big Data Analytics in Traffic and Transportation Engineering:

Emerging Research and Opportunities

Sara Moridpour
RMIT University, Australia

Alireza Toran Pour
RMIT University, Australia

Tayebeh Saghapour
RMIT University, Australia

A volume in the Advances in Civil and Industrial Engineering (ACIE) Book Series

IGI Global
DISSEMINATOR OF KNOWLEDGE

Published in the United States of America by
IGI Global
Engineering Science Reference (an imprint of IGI Global)
701 E. Chocolate Avenue
Hershey PA, USA 17033
Tel: 717-533-8845
Fax: 717-533-8661
E-mail: cust@igi-global.com
Web site: http://www.igi-global.com

Library of Congress Cataloging-in-Publication Data

Names: Moridpour, Sara, 1980- author.
Title: Big data analytics in traffic and transportation engineering : emerging research and opportunities / by Sara Moridpour.
Description: Hershey, PA : Engineering Science Reference, [2019]
Identifiers: LCCN 2018038725| ISBN 9781522579434 (hardcover) | ISBN 9781522579441 (ebook)
Subjects: LCSH: Pedestrian accidents--Australia--Melbourne (Vic.)--Statistical methods. | Traffic accidents--Australia--Melbourne (Vic.)--Statistical methods. | Roads--Interchanges and intersections--Safety measures--Statistical methods. | Big data--Australia--Melbourne (Vic.)
Classification: LCC HE5614.5.A8 M67 2019 | DDC 363.12/5--dc23 LC record available at https://lccn.loc.gov/2018038725

This book is published in the IGI Global book series Advances in Civil and Industrial Engineering (ACIE) (ISSN: 2326-6139; eISSN: 2326-6155)

British Cataloguing in Publication Data
A Cataloguing in Publication record for this book is available from the British Library.

Advances in Civil and Industrial Engineering (ACIE) Book Series

ISSN:2326-6139
EISSN:2326-6155

Editor-in-Chief: Ioan Constantin Dima, University Valahia of Târgovişte, Romania

MISSION

Private and public sector infrastructures begin to age, or require change in the face of developing technologies, the fields of civil and industrial engineering have become increasingly important as a method to mitigate and manage these changes. As governments and the public at large begin to grapple with climate change and growing populations, civil engineering has become more interdisciplinary and the need for publications that discuss the rapid changes and advancements in the field have become more in-demand. Additionally, private corporations and companies are facing similar changes and challenges, with the pressure for new and innovative methods being placed on those involved in industrial engineering.

The **Advances in Civil and Industrial Engineering (ACIE) Book Series** aims to present research and methodology that will provide solutions and discussions to meet such needs. The latest methodologies, applications, tools, and analysis will be published through the books included in **ACIE** in order to keep the available research in civil and industrial engineering as current and timely as possible.

COVERAGE

- Quality Engineering
- Construction Engineering
- Engineering Economics
- Optimization Techniques
- Ergonomics
- Earthquake engineering
- Materials Management
- Hydraulic Engineering
- Productivity
- Urban Engineering

IGI Global is currently accepting manuscripts for publication within this series. To submit a proposal for a volume in this series, please contact our Acquisition Editors at Acquisitions@igi-global.com or visit: http://www.igi-global.com/publish/.

The Advances in Civil and Industrial Engineering (ACIE) Book Series (ISSN 2326-6139) is published by IGI Global, 701 E. Chocolate Avenue, Hershey, PA 17033-1240, USA, www.igi-global.com. This series is composed of titles available for purchase individually; each title is edited to be contextually exclusive from any other title within the series. For pricing and ordering information please visit http://www.igi-global.com/book-series/advances-civil-industrial-engineering/73673. Postmaster: Send all address changes to above address.

Titles in this Series

For a list of additional titles in this series, please visit:
https://www.igi-global.com/book-series/advances-civil-industrial-engineering/73673

Optimization of Design for Better Structural Capacity
Mourad Belgasmia (Setif 1 University, Algeria)
Engineering Science Reference • ©2019 • 283pp • H/C (ISBN: 9781522570592) • US $215.00

Reusable and Sustainable Building Materials in Modern Architecture
Gülşah Koç (Yildiz Technical University, Turkey) and Bryan Christiansen (Global Research Society, LLC, USA)
Engineering Science Reference • ©2019 • 302pp • H/C (ISBN: 9781522569954) • US $195.00

Measuring Maturity in Complex Engineering Projects
João Carlos Araújo da Silva Neto (Magnesita SA, Brazil) Ítalo Coutinho (Saletto Engenharia de Serviços, Brazil) Gustavo Teixeira (PM BASIS, Brazil) and Alexandro Avila de Moura (Paranapanema SA, Brazil)
Engineering Science Reference • ©2019 • 277pp • H/C (ISBN: 9781522558644) • US $245.00

Big Data Analytics for Smart and Connected Cities
Nilanjan Dey (Techno India College of Technology, India) and Sharvari Tamane (Jawaharlal Nehru Engineering College, India)
Engineering Science Reference • ©2019 • 348pp • H/C (ISBN: 9781522562078) • US $225.00

Contemporary Strategies and Approaches in 3-D Information Modeling
Bimal Kumar (Glasgow Caledonian University, UK)
Engineering Science Reference • ©2018 • 313pp • H/C (ISBN: 9781522556251) • US $205.00

New Approaches, Methods, and Tools in Urban E-Planning
Carlos Nunes Silva (University of Lisbon, Portugal)
Engineering Science Reference • ©2018 • 407pp • H/C (ISBN: 9781522559993) • US $205.00

For an entire list of titles in this series, please visit:
https://www.igi-global.com/book-series/advances-civil-industrial-engineering/73673

IGI Global
DISSEMINATOR OF KNOWLEDGE

701 East Chocolate Avenue, Hershey, PA 17033, USA
Tel: 717-533-8845 x100 • Fax: 717-533-8661
E-Mail: cust@igi-global.com • www.igi-global.com

Table of Contents

Preface

A growing number of researchers have recently focused on improving the sustainability of transportation systems by converting routine motorised travel into active modes of transport. The importance of physical activity and its impact on health has not only attracted the attention of practitioners, but it has also turned the attention of planners and policy makers to the achievement of sustainable transportation by enhancing active travel behaviour. To identify effective strategies for increasing pedestrian and bicycle transportation in a specific local area, planners need to identify how the current levels of accessibility in neighbourhoods affect transport mode choice. Although many studies have been conducted on modelling active transportation, the importance of accessibility has been neglected. Therefore, this book proposes new approaches to the measurement of walking, cycling and public transport accessibility while using new measurements in regression models to examine how accessibility can affect active transportation. Promoting active transportation requires better accessibility to activities and places of interest. Hence, in the first step, recognition of the level of accessibility in neighbourhoods is essential. Several approaches have been developed and used in the research literature which measure accessibility for non-motorised modes of transport. However, existing measurements have some limitations that may affect the accuracy of accessibility levels. Therefore, the present book focuses on the development of new accessibility measures for public transport, walking and cycling, which overcome the limitations of past measures.

With respect to the public transport accessibility index, in existing approaches, the distribution of the population is ignored. Therefore, this book proposes a new method of measurement which extends two common approaches incorporating population density. This book also introduces a new index for measuring cycling accessibility, which is a gravity-based measure. Whilst existing cycling accessibility measures are dependent on travel data, this new index measures levels of accessibility independently of travel data. Regarding walking accessibility, existing methods use travel distance or land-use features to measure walkability. However, the method proposed in this book not only considers walking distance thresholds, but also incorporates the diversity and intensity of land use. This book also focuses on the application of accessibility measures and the importance of considering accessibility as the explanatory variable in modelling active transportation. For this purpose, new measurements are employed in regression models versus land use factors to examine the performance as well as the importance of including accessibility measures in transport modelling.

In the Melbourne metropolitan area in Australia, an average of 34 pedestrians were killed in traffic accidents every year between 2004 and 2013, and vehicle-pedestrian crashes accounted for 24% of all fatal crashes. Mid-block crashes accounted for 46% of the total pedestrian crashes in the Melbourne metropolitan area and 49% of the pedestrian fatalities occurred at mid-blocks. Many studies have examined factors contributing to the frequency and severity of vehicle-pedestrian crashes. While many of the studies have chosen to focus on crashes at intersections, few studies have focussed on vehicle-pedestrian crashes at mid-blocks. Since the factors contributing to vehicle crashes at intersections and mid-blocks are significantly different, more research needs to be done to develop a model for vehicle-pedestrian crashes at mid-blocks. Furthermore, socioeconomic factors are known to be contributing factors to vehicle-pedestrian crashes. Although several studies have examined the socioeconomic factors related to the locations of crashes, few studies have considered the socioeconomic factors of the neighbourhoods where road users live in vehicle-pedestrian crash modelling. In vehicle-pedestrian crashes in the Melbourne metropolitan area 20% of pedestrians, 11% of drivers and only 6% of both drivers and pedestrians had the same postcode for the crash and residency locations. Therefore, an examination of the influence of socioeconomic factors of their neighbourhoods, and their relative importance will contribute to advancing knowledge in the field, as very limited research has been conducted on the influence of socioeconomic factors of both the neighbourhoods where crashes occur and where pedestrians live.

In order to identify factors contributing to the severity of vehicle-pedestrian crashes, three models using different decision trees (DTs) have been proposed in the current book. To improve the accuracy, stability and robustness of the DTs, bagging and boosting techniques have been used. The results show that the boosting technique improves the accuracy of individual DT models by 46%. Moreover, the results of boosting DTs (BDTs) show that neighbourhood social characteristics are as important as traffic and infrastructure variables in influencing the severity of pedestrian crashes. In this book, neighbourhood factors associated with road users' residents and location of crash are investigated using BDT model. Furthermore, partial dependence plots are applied to illustrate the interactions between these factors. It has been found that socioeconomic factors account for 60% of the 20 top contributing factors to vehicle-pedestrian crashes. This paper reveals that socioeconomic factors of the neighbourhoods where road users live and where crashes occur are important in determining the severity of crashes, with the former having a greater influence. Hence, road safety counter-measures, especially those focussing on road users, should be targeted at these high-risk neighbourhoods.

To develop effective and targeted safety programs, the location and time-specific influences on vehicle-pedestrian crashes must be assessed. Therefore, spatial autocorrelation has been applied in this book for the examination of vehicle-pedestrian crashes in geographic information systems (GISs) to identify any dependency between time and location of these crashes. Spider plotting and Kernel Density Estimation (KDE) have been then used to determine the temporal and spatial patterns of vehicle-pedestrian crashes for different age groups and gender types. Temporal analysis shows that pedestrian age has a significant influence on the temporal distribution of vehicle-pedestrian crashes. Furthermore, men and women have different crash patterns. In addition, the results of the spatial analysis show that areas with high risk of vehicle-pedestrian crashes can vary during different times of the day for different age groups and gender types.

The book is organised into seven chapters. A brief description of each of the chapters are as follows.

Chapter 1 has used the large dataset of Victorian Integrated Survey of Travel and Activity (VISTA) to introduce a new approach measuring public transport accessibility within the Melbourne region, Australia, as the case study. A Public Transport Accessibility Index (PTAI) is a combined measure of public transport service frequency and population density as an important distributional indicator. The proposed index is compared with two common existing approaches using regression models.

In Chapter 2, a new index for measuring bikeability in metropolitan areas is presented. The Cycling Accessibility Index (CAI) has been developed for computing cycling accessibility within Melbourne metropolitan, Australia. The CAI is defined consistent with gravity-based measures of accessibility. This index measures cycling accessibility levels considering mixed use developments as well as travel distance between origins and destinations.

Chapter 3 presents a new approach for measuring walkability within Melbourne region, Australia. An integrated approach combining transport and land-use planning concepts has been employed to construct the Walking Access Index (WAI), which is a location-based measure for accessibility.

In Chapter 4, three models using different Decision Trees (DTs) have been developed to identify factors contributing to the severity of vehicle-pedestrian crashes. To improve the accuracy, stability, and robustness of the DTs, bagging and boosting techniques have been used in this chapter.

In Chapter 5 neighbourhood factors associated with road users' residents and location of crash are investigated using BDT model. Furthermore, partial dependence plots are applied to illustrate the interactions between these factors.

In Chapter 6, spatial autocorrelation has been applied for the examination of vehicle-pedestrian crashes in Geographic Information Systems (GISs) to identify any dependency between time and location of these crashes. Spider plotting and Kernel Density Estimation (KDE) have been then used to determine the temporal and spatial patterns of vehicle-pedestrian crashes for different age groups and gender types.

Chapter 7 aims to identify contributing factors on vehicle-pedestrian crash severity of pedestrians with less than 18 years of age or school-aged pedestrians. Reasonable walking distance to schools is applied in Geographic Information Systems (GIS) to identify vehicle-pedestrian crashes around schools. Then Boosted Decision Tree (BDT) and Cross-Validation (CV) technique are applied to explore the significant factors.

Chapter 1
Measuring Public Transport Accessibility in Metropolitan Area

ABSTRACT

Improving access to public transport can be considered an effective way of reducing the negative side-effects of motorised commuting. This chapter used the large dataset of Victorian Integrated Survey of Travel and Activity (VISTA) to introduce a new approach measuring public transport accessibility within the Melbourne region, Australia. A public transport accessibility index (PTAI) is a combined measure of public transport service frequency and population density as an important distributional indicator. Although many studies have measured access levels to public transport stops/stations, there has been limited research on accessibility that integrates population density within geographical areas. Employing geographical information system (GIS), a consistent method is introduced for evaluating public transport accessibility for different levels of analysis, from single elements, including public mode stops, to network analysis. The proposed index is compared with two common existing approaches using regression models. Key findings indicate that the PTAI has a stronger association whilst showing more use of public transport in areas with higher values of the PTAI.

DOI: 10.4018/978-1-5225-7943-4.ch001

1.1 INTRODUCTION

Shifting from private motorised vehicles to non-motorised modes of transport such as public transportation, walking and cycling can increase the sustainability of transportation and consequently, improve the environment, public health and the economy (Elias and Shiftan, 2012). A user-friendly public transportation system should consider accessibility to stops/stations, the mobility of the system and the connectivity to other transportation modes (Cheng and Chen, 2015). In recent decades, automobile-oriented developments along with increased car ownership have encouraged people to spend more time travelling by car. High levels of car dependency not only affect the quality of life, but also threaten people's health. On the other hand, growing use of private motorization has resulted in critical issues such as traffic congestion and environmental impacts. Use of public transport has been recently considered within the definition of active transport as it often involves some walking or cycling to make connections from the origins to the destinations (Taniguchi et al., 2013). For this reason, the provision of high levels of accessibility for public transport systems with good connectivity can promote active transport and sustainability.

Australia has been categorized as a country with high car ownership (Lucas, 2012) with particular groups of people such as youths, seniors, low-income households and Aboriginals encountering difficulties in accessing work, education and social or cultural activities (Lucas, 2012, Altman and Hinkson, 2007, Johnson et al., 2011). It has been shown that some suburban and regional areas in Australia are disadvantaged with respect to public transport, where distance is a major barrier (Currie and Stanley, 2007). As Wang and Chen (2015) argued transportation equity affects residents' economic as well as social opportunities. In other words, transportation problems may result in social exclusion, as reported in several studies (Fransen et al., 2015, Priya and Uteng, 2009, Delmelle and Casas, 2012, Lucas, 2011).

Ceder et al. (2009) argued that an effective public transport service can be defined as minimum in-vehicle travel time and waiting time. Although physical access to public transport stops is important, the time taken to travel between an origin and destination by public transport modes can be considered as another substantial factor (Lei and Church, 2010). Accessibility measures have been generally categorized into three groups, access to public transport stops, duration of journeys by public transport modes and access to destinations by public transport modes (Mavoa et al., 2012). A large number of studies have focused on the proximity to a public transport stop/station for measuring accessibility (Biba et al., 2010, Currie, 2010, Furth et al., 2007, Lovett et al., 2002). Typically, the maximum acceptable walking distance is considered as 400 m and 800 m for public transport stops or stations (Currie, 2010, Currie, 2004, El-Geneidy et al., 2010).

Along with studies that focus on access to public transport stops, some studies focus on the duration of a journey undertaken by public transport modes (O'Sullivan et al., 2000, Benenson et al., 2011). O'Sullivan et al. (O'Sullivan et al., 2000) measured public transport accessibility generating maps of accessible areas with the same travel time. In another study, Cheng and Agrawal (Cheng and Agrawal, 2010) introduced an accessibility measurement tool which calculates a public transport service area considering travel time. Yigitcanlar et al. (Yigitcanlar et al., 2007) introduced a GIS-based land use and public transport accessibility index (LUPTAI). This approach measures accessibility based on both public transport travel time and walking distances utilizing GIS analysis techniques. They used an origin-based accessibility and destination-based GIS technique and applied the index to two pilot studies in the Gold Coast, Australia. Their findings indicated that the LUPTAI could easily be applied to a range of different of land use categories. Access to a destination using public transport modes is another technique of measuring accessibility (Curtis and Scheurer, 2010). Huang and Wei (Huang and Wei, 2002) measured access via public transport using business and industrial land parcels. They computed the distance between census tracks, as the origin points, and those parcels using a public transport network.

Service frequency is a critical aspect of accessibility, which varies in different commuting times (Mavoa et al., 2012). Several studies conducted using service frequency as a complement in their approach or as an independent measure. Service frequency-based measurements have been categorized into two general groups Mavoa et al. (Mavoa et al., 2012). For the first group, a minimum service frequency standard has been adopted. This approach excludes the public transport that do not meet the standard (Curtis and Scheurer, 2010). The second group includes all public transport stops while using service frequency. For instance, using the number of trips per week for each stop or station (Currie, 2010) or category, the service frequency is measured by how often a public transport mode arrives (Yigitcanlar et al., 2007). A needs-gap approach used by Currie (Currie, 2004) identified spatial gaps in terms of public transport supply in Hobart, Australia. A more recent version of that approach was developed for metropolitan Melbourne (Currie, 2010). These studies used a combined measure of service frequency and access distance, which was calculated for each census collector district (CCD). Among a series of service frequency methodological developments within this area, the PTAL (Public Transport Accessibility Level) is a UK approach which measures the level of accessibility. The PTAL provides a six-level rating scale of public transport accessibility, which includes measures such as access walk time, service frequency and waiting time. This approach calculates the level of access by public transport to points of interest (Wu and Hine, 2003, Currie, 2010).

Numerous studies have focused on measuring public transport accessibility. However, there has been limited work considering the distribution of the population in measuring accessibility levels. We present a new index to measure public transport accessibility and describe its application to increase understanding of public transport usage in metropolitan Melbourne, Australia. There is a need to incorporate different frequencies of public transport modes, public transport routes and population densities in measuring public transport accessibility. In this chapter a new index is presented that can be used to classify levels of public transport accessibility. The method has been applied to the Melbourne metropolitan area, which is served by a public transport system that includes train, tram and bus services. The following section introduces the methodology, and Section 3.3 describes the computation of the index. An analysis and the results of the application of the PTAI in the Melbourne region, along with a comparison of the results between the new index and existing approaches, are presented in Section 3.4. Section 3.5 discusses the results, while Section 3.6 summarizes the findings and outlines avenues for future research.

1.2 METHODOLOGY

The aim of this study was to develop an index for the measurement of the level of accessibility to public transport in Melbourne's 9510 Statistical Areas level 1 (SA1s)[1], the second smallest geographic area defined in the Australian Statistical Geography Standard (ABS, 2011b). According to the Australian Government Department of Health and Ageing (Neighbourhood Planning and Design, 2009), the physical characteristics of neighbourhoods are accessible based on walkable catchments. This is generally defined as 5 to 10 minutes walking to/from public transport stops/stations. SA1 districts were found to have the closest conformity to walking catchments. In order to define the index, two factors, a weighted equivalent frequency (WEF) and the ratio of population density in SA1s and buffer areas (service areas of different public transport modes) are calculated. This work is consistent with Lei and Church's (Lei and Church, 2010) classification, as it deals with physical access to public transport stops/stations in terms of walking time and service frequency. Furthermore, the work fits into the first category, access to public transport stops, of the more general three-way classification scheme developed by Mavoa et al. (2012). The methodology has been developed for metropolitan Melbourne, where areas with a denser public transport network and population show greater access to all nearby destinations. The databases and the study area, the conceptual framework, and existing methods and approaches are presented in this section to describe the process for calculating the index.

1.2.1 Datasets and Study Areas

For calculating the PTAI three main sets of databases was adopted as follows:

- Special database of public transport stops/stations including public buses, trams stops and train stations. This dataset was obtained from the Victorian Government open data sources (ckan, 2016). This data contains approximately 17800 bus stops, 1700 tram stops, and 240 train stations within the Melbourne region (see Figure 1. for the distribution of stops/station within the Melbourne region).
- Service frequency data were calculated from the timetables for each mode during the morning peak hours (7 to 9 am). For example, for a bus route with average 20-minute services during the peak hours, the frequency was calculated to be 3. Timetables are accessible on the Public Transport Victoria (PTV) website (https://www.ptv.vic.gov.au).
- A database of points of interest (POIs) was obtained from the Australian Urban Research Infrastructure Network (AURIN, 2016). This included urban centres, significant buildings, landmarks, public spaces, community facilities and indigenous locations, consisting of 15588 points.

A spatial database of statistical areas from the 2011 Census for the Melbourne Region which was obtained from the Australian Bureau of Statistics (ABS, 2011). This data set contains the total usual resident population and total number of dwellings from the 2011 Census of Population and Housing for mesh blocks (the smallest geographical unit released by the ABS) and all other statistical areas, including SA1s. According to the ABS, the Melbourne region contains 53074 mesh blocks, 9510 SA1s, 277 statistical areas level 2 (SA2s) and 31 local government areas (LGAs). Mesh blocks are the smallest geographical unit released by the ABS and all other statistical areas are built up from or, approximated by, whole mesh blocks.

1.2.2 Conceptual Framework

The PTAI consists of two main procedures. The first step relates to the POIs and public transport services, and the second step involves calculating the population density in both walking catchments and SA1s. Figure 2. shows the conceptual framework of the calculation process for the PTAI. For a given POI, the shortest distance to a public transport stop/station is defined. Thereafter, the equivalent frequency is computed following the steps shown. On the other side, as shown, for public transport modes' service areas, the proportion of population density is calculated for each buffer area and SA1.

Figure 1. Distribution of public transport stops/stations in metropolitan Melbourne

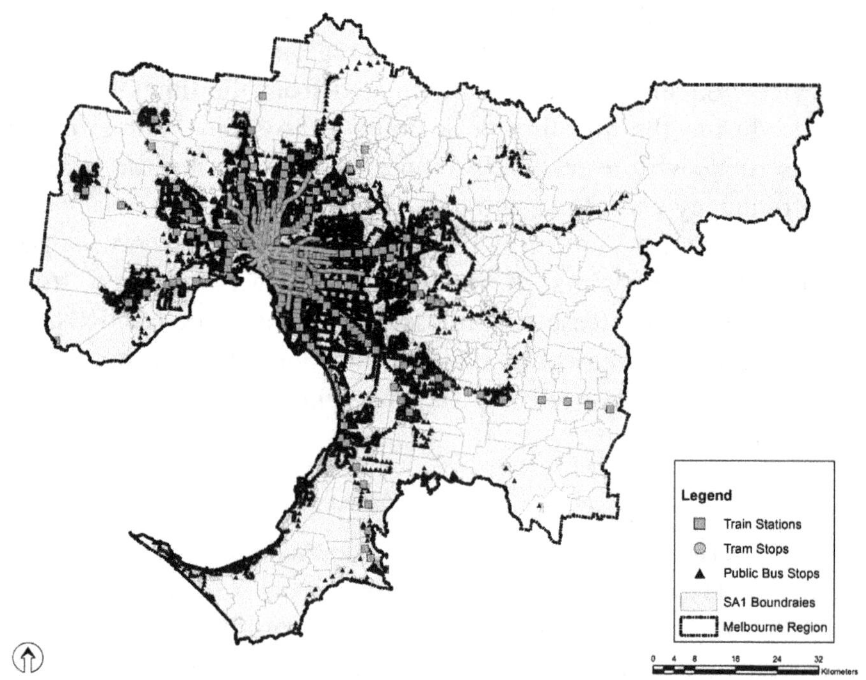

Figure 2. Conceptual framework of the calculation process

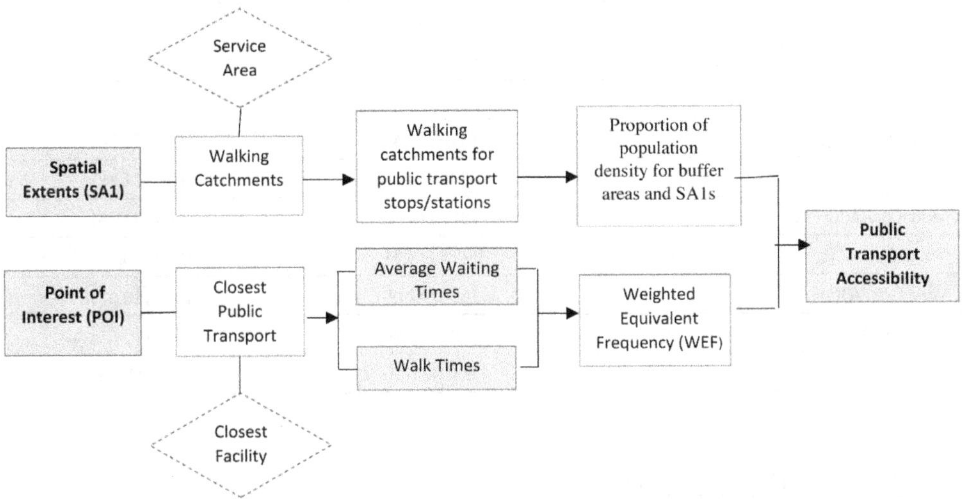

1.2.3 Approach

The approach introduced here extends two common approaches, including the UK approach (London, 2010) measuring public transport accessibility levels (PTAL) and the supply index (SI) introduced by Currie (2010). PTAL rates public transport service access using a six-level scale, and includes measures such as walk time, waiting time and service frequency. This index calculates the sum of equivalent doorstep frequency (EDF) of all different public transport modes. SI is a supply index calculated for Melbourne's 5839 census collector districts (CCDs). The index is a combined measure of service frequency (number of public transport vehicle arrivals per week) and access distance, as shown in Equation 1.

$$SI_{CCD} = \sum N \left(\frac{Area_{B_n}}{Area_{CCD}} * SL_{B_n} \right) \tag{1}$$

where, SI_{CCD} is the supply index for CCDs and N is the number of walking buffers to public transport stops/stations in each CCD. B_n is the buffer n for each stop/station, Area is the square kilometre area of the CCD and SL is the service level of the public transport modes (Currie, 2010). PTAL and SI were genetared for SA1s and the results are presented in Table 1.

Table 1. Public transport accessibility levels (PTAL) and Supply index (SI) For SA1s

PTAL/SI Categories	PTAL	SI
	No. (%) of SA1s	No. (%) of SA1s
Zero Access/Supply	52 (0.55)	267 (2.81)
Very Poor/Very Low	1370 (14.41)	2117 (22.26)
Poor/Low	1398 (14.70)	2014 (21.18)
Moderate/Below Average	1857 (19.53)	2069 (21.76)
Good/Above Average	1415 (14.88)	1032 (10.85)
Very Good/High	1624 (17.08)	1000 (10.52)
Excellent/Very High	1794 (18.86)	1011 (10.63)
Total	9510 (100)	9510 (100)

1.2.2 PTAI Calculation

The Public Transport Accessibility Index (PTAI) calculation includes two parts. First part is related to find the closest public transport stop/station to a given POI and the second part relates to calculating the population density ratio (Saghapour, Moridpour, & Thompson, 2016). As mentioned previously, there are approximately 20,000 public transport stops/stations within the Melbourne region. This area has about 16,000 POIs including community services and facilities, landmarks, non-residential and public buildings. Using ArcGIS, Closest Facility from the network analyst tools was applied separately for each public transport mode. For instance, considering a shopping centre as a POI, the distance of the nearest public bus stop was measured. Same process was applied for the closest tram stop and train station. The following sections describe the first part of the index formulation. The calculation of the WEF extends the approach used in measuring public transport accessibility levels in London (London, 2010).

Walk Time (WT)

First step is calculating the WT from a given POI to the closest public transport stop/station. Walk distances calculated using ArcGIS network analysis. Thereafter, walk distances were converted to a measure of time, assuming an average walk speed of 4.8 kilometres/hour or 80 metres/minute (London, 2010). The maximum walk time of 10 minutes (800 metres) was defined for buses and trams stops, 15 minutes (1200 metres) for train stations.

Average Waiting Time (AWT)

For each selected route, the AWT was defined as the interval between services. For instance, for a public transport mode running services every 5 minutes, the AWT is 2.5 minutes. In other words, a passenger may have to wait about 6 minutes for the arrival of a desired service. The AWT is estimated as half the headway (i.e. the time interval between services), as shown in Equation (2).

$$AWT_{ij} = 0.5 * (60 / F_{ij}); i = 1,2,3,\ldots,n; j = 1,2,3 \quad (2)$$

where, AWT_{ij} is the average waiting time (in minutes) at the closest stop/station to the POI i for public transport mode j and F_{ij} is the frequency of mode j (defined as the number of services per hour) at the closest stop/station to the POI i.

Total Access Time (TAT)

Summed value of WT and AWT gives us the Total Access Time (TAT) of a selected *POI* to the nearest public transport stop/station (see Equation 3).

$$TAT_{ij} = WT_{ij} + AWT_{ij}; i = 1,2,3,\ldots,n; j = 1,2,3 \quad (3)$$

where, TAT_{ij} is the total access time (minutes) of public transport mode j at the closet stop/station to the POI I, and WT_{ij}, as explained above, is the walk time (in minutes) from the POI i to the closest stop/station of public transport mode j.

Equivalent Frequency (EF)

TATs were converted to an equivalent frequency using Equation (4). This measures the doorstep availability of a route at the specified *POI*. The Equivalent Frequency (*EF*) as presented in Equation (5) is calculated as 30 minutes divided by the *TAT*.

$$EF_{ij} = \frac{30}{TAT_{ij}}; i = 1,2,3,\ldots,n; j = 1,2,3 \quad (4)$$

where, EF_{ij} is the equivalent frequency for public transport mode j at the closet stop/station to the POI i.

Weighted Equivalent Frequency (WEF)

The Weighted Equivalent Frequency (*WEF*) is calculated as a summation of the *EF*s of public transport modes with a weighting in favour of the most dominant mode (Equation 5).

$$WEF_{ij} = \alpha EF_{id} + \beta \sum_{i} \sum_{j \neq d} EF_{ij}; i = 1, 2, \ldots, n; j = 1, 2, 3 \quad (5)$$

WEF_{ij} is the weighted equivalent frequency for public transport mode j at the closest stop/station to the POI i, EF_{id} is the equivalent frequency of the most dominant public transport mode at the closest stop/station to the POI i, α and β are the coefficients considered for the equivalent frequency of the most dominant public transport mode and all other public transport modes. Considering factors of popularity, time and number of passengers transferred by public transport modes, α and β were defined as 1 for the train (the dominant mode) and 0.5 for the two other modes.

WEFs for SA1s

The WEFs calculated for POIs were transferred to the SA1s. For this purpose, spatial joining (using ArcGIS 10.2) was used based on the criterion of closeness to the boundaries of SA1s. Hence, considering any POI, the WEF was transferred from the one which had the minimum distance to the boundary of its surrounding SA1s. The reason for this was that since SA1 boundaries are compatible with roads, the closest POI to a SA1 boundary also has the shortest distance to the road. This may make particular POIs more accessible than their counterparts.

Population Density

The second part of the PTAI calculation is computing the population density ratio. Population densities were calculated for both buffer areas and SA1s. Buffer areas of 400 metres were considered bus and tram stops and 800 metres were assumed for train stations. Thereafter, buffer areas were overlapped with SA1s, using to calculate the share of population density for each SA1. Populations within buffer areas were calculated based on the proportion of buffer areas overlapping the mesh blocks (assuming a homogenous distribution of populations within mesh blocks).

For each SA1 the PTAI is calculated using the formula given in Equation (6). The index is a combined measure of *WEF* and population density ratio given as:

if $D_{B_{ij}} = 0$; (6)

$$PTAI_{SA1} = \sum_{j=1}^{3}\sum_{i=1}^{I}\left(1 + \frac{D_{B_{ij}}}{D_{SA1_i}}\right) * WEF_{SA1_i}$$

if $D_{B_{ij}} \neq 0$;

$$PTAI_{SA1} = \sum_{j=1}^{3}\sum_{i=1}^{I}\left(\frac{D_{B_{ij}}}{D_{SA1_i}}\right) * WEF_{SA1_i}$$

where, $PTAI_{SA1}$ denotes the public transport accessibility index for a given *SA1* and D_{Bij} is the population density of buffer *i* for public transport mode *j*, D_{SA1} is the population density of the *SA1*, and WEF_{SA1} is the weighted equivalent frequency calculated for the corresponding *SA1*. The index counts the overlapping buffer areas. For instance, where there is a place within possible walking distance to both bus and tram stops, the measurements are double-counted, which indicates that those areas have a higher level of accessibility to public transport. A higher value of the PTAI indicates a higher level of accessibility. A value of 0 indicates that there is either no accessibility or no population in a given SA1. In areas with no population or non-residential uses, the PTAI is equal to $\mathrm{WEF}_{\mathrm{SA1}}$. More details on PTAI calculation is provided is studies by Saghapour et al. (Saghapour et al., 2016c, Saghapour et al., 2016b).

Table 2 presents the ranges and categories of the PTAI. The index was grouped into six main categories including very poor, poor, moderate, good, very good and excellent plus a zero group. The classification method used for

PTAI categories is based on Quantile, since they are one of the best methods for simplifying the map-reading (Brewer and Pickle, 2002). Zero accessibility is provided for 16243 residents or 0.55% of the total population. Very poor areas are mostly located in outer Melbourne. Overall, around 50% of the total population have zero to moderate accessibility to public transport.

Figure 3 illustrates the distribution of PTAI categories in the Melbourne region. As explained above, the PTAI is categorized into 6 bands. The first category represents very poor accessibility, while the last category corresponds to an excellent level of accessibility to public transport. The first category has been further sub-divided into sub-levels to provide better clarity. High levels of accessibility from good to excellent are mostly concentrated in the inner parts of the Melbourne region. As the figure shows, outer Melbourne, where public transport is mainly provided by public buses has low levels of accessibility in comparison to inner parts and the CBD.

Table 3 presents a summary of the descriptive statistics of the index components. This shows that there are on average 414 residents in each SA1 with an average area of 0.93 km^2. The average number of stops/stations per SA1 is 2.1, which receive a total of 9.6 services during peak times. The average WEF per SA1 is 5.5 and the average value of the PTAI per SA1 is 8.8. On average, 28% of the Melbourne area is covered by the walking catchments of bus stops. This proportion is 4% and 3% for train stations and tram stop walking buffers, respectively.

Table 2. PTAI ranges and categories

Ranges	PTAI Categories	SA1s	Population
		No. (%)	No. (%)
0	N/A	52 (0.55)	16243 (0.41)
< 2	Very Poor	1331 (14.00)	538536 (13.66)
2 – 3.5	Poor	1607 (16.90)	671449 (17.04)
3.5 - 6	Moderate	1791 (18.83)	751327 (19.06)
6 - 12	Good	1969 (20.70)	801520 (20.34)
12 - 20	Very Good	1480 (15.56)	623111 (15.81)
> 20	Excellent	1280 (13.46)	539025 (13.68)
Total	N/A	9510 (100.00)	3941211 (100.00)

Figure 3. Distribution of PTAI categories in the Melbourne region

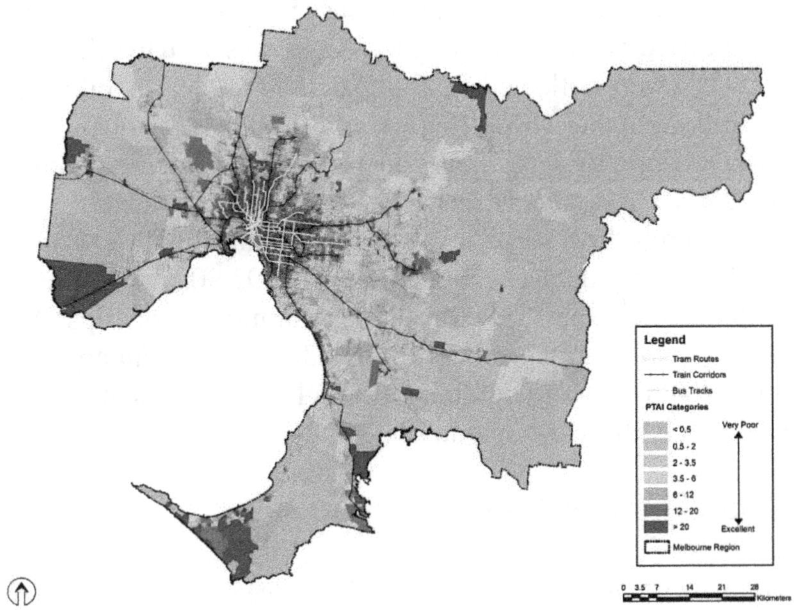

Table 3. Descriptive statistics of indicators in each SA1

Indicators	Mean	S.D.
Area (km2)	0.93	10.2
Population	414	209.5
Frequency of Bus services	2.2	1.5
Frequency of Tram services	2.9	4.1
Frequency of Train services	4.5	2.6
Number of public transport stops/stations per SA1	2.1	2.5
WEF	5.5	5.3
PTAI	8.8	10.7

1.3 DATA ANALYSIS

The Victorian Integrated Survey of Travel and Activity (VISTA) data set was adopted to assess and evaluate the index. The VISTA dataset was published by the Department of Economic Development, Jobs, Transport and Resources (EDJTR) in 2009. The VISTA is a cross-sectional survey conducted from 2009 until July 2010. It covers the Melbourne Statistical Division (MSD), as defined by the Australian Bureau of Statistics (ABS), and the regional cities of Geelong, Ballarat, Bendigo and Shepparton, and the Latrobe Valley. A stratified random sampling technique was used to select residential properties. Data were collected regarding demographic, trip information and car ownership. A total of 16411 households (42,002 individuals) responded, with a response rate of 47%. This paper only considered responses within the MSD (22,201 individuals). The VISTA recorded travel in the form of trip stages, where a "trip stage" is a segment of travel with a single purpose and mode. Hence, the dataset contains details of 93,902 trips stages made by 22,184 individuals in the MSD. From the total number of trips 18701 trips were made by public transport modes including train, tram and public bus. Whilst the VISTA dataset contained the SA1 codes, the statistical analysis was applied using the same spatial scale.

1.3.1 Modelling and Interpretation

Built environment factors, as well as public transport access measurements, were combined with the VISTA dataset using the SA1 codes. The VISTA dataset contains trip record information for 22,184 individuals who were randomly selected from 1,822 SA1s. The following sections present the results of the models applied to the data while comparing the new index with the previous measurements.

Models for this study were estimated using NBR techniques, which are able to use count data and require non-negative integers for the count dependent variable. Since the number of trips is always a positive integer, this study adopted the NBR model (Coruh et al., 2015; Saghapour, Moridpour, & Thompson, 2017).

NBR models were used to analyse the effects of explanatory variables on the number of public transport trips. Linear regression techniques have been widely used to examine travel behaviour (Krizek, 2003, Kitamura et al., 1997, Frank and Pivo, 1994). However, linear regression analysis requires the models' residuals to follow the normal distribution (Nachtsheim et al., 2004), while distributions of trip frequencies are often positively skewed, and deviate from the normality assumption (Cao et al., 2006).

In Poisson regression, it is assumed that the dependent variable Y (the frequency of walking trips in this study) is Poisson-distributed given the explanatory variables X1, X2,..., Xp. This means that the probability of observing Y= k trips, can be obtained by the Poisson distribution function (Cao et al., 2006):

$$P\left(Y = K \mid X_1, X_2, \ldots, X_p\right) = \frac{e^{-\lambda}\lambda^k}{k!}, k = 0,1,2,3,\ldots, \tag{7}$$

where, the conditional mean k is an exponential function of the explanatory variables. That is,

$$\lambda = \exp\left(\beta_0 + \beta_1 X_1 + \beta_2 X_2 + \ldots + \beta_p X_p\right) \tag{8}$$

where, the fitted value of Y for the ith case, $\hat{Y}_i\left(i = 1,2,\ldots,N\right)$, is denoted $\hat{\lambda}_i$.

Poisson regression assumes equality of mean variances. However, this assumption is frequently violated in empirical data. As shown in Figure 4, there is some evidence of over-dispersion (variance > mean) in active trips. Alternatively, the NBR model captures the over-dispersion effect by introducing an unobserved effect into the conditional mean, λ, of the Poisson model:

$$\lambda = \exp\left(\beta_0 + \beta_1 X_1 + \beta_2 X_2 + \ldots + \beta_p X_p + \varepsilon\right) \tag{9}$$

where, exp(ε) has a gamma distribution with mean 1 and variance α (also called the dispersion parameter). Poisson regression is a special case of NBR in which α equals 0. As shown if Figure 4. public transport trips used in this study follow the assumed pattern (Mean < Variance).

M1 presents the results of NBR model considering all the predictor variables and the PTAI as the public transport accessibility measure. M2 and M3 contain all variables used in the M1; however, SI and PTAL are used for public transport accessibility measures, respectively. Public transport trips are defined as a count-dependent variable. Age, gender, car licence, employment type, household size, household structure, number of years lived at the same address and the number of cars in the HH were used as socioeconomic variables (Lee et al., 2014, Jun et al., 2012, Shay and Khattak, 2012, Ewing and Cervero, 2010, Winters et al., 2010, Engelfriet and Koomen, 2017). Table 5 shows the list of independent variables and their descriptions. Built environment factors include roadway measure (RM), land-use mix entropy index (EI) and public transport accessibility measurements (PTAI/SI/PTAL). The RM examines how far the network spreads over an area. It is quantified by the

Figure 4. Histogram of public transport

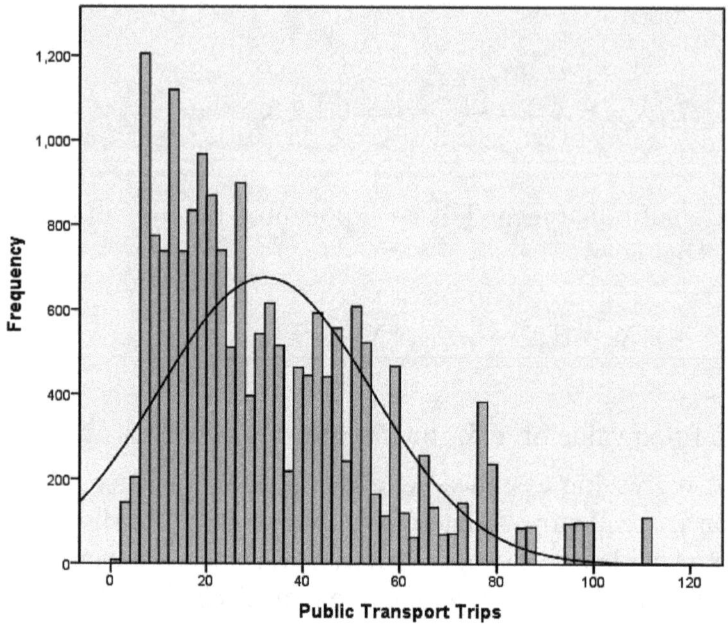

total roadway length divided by the total area and the distance is normalised by $100m^2$. The EI was calculated using Equation (10) (Lee et al., 2014). The values vary from 0 to 1, while 1 indicates a perfect balance among different types of land uses and 0 represents homogeneity.

$$EI = -\left(\sum_{j=1}^{J} \frac{P_j \cdot \ln P_j}{\ln J} \right) \quad (10)$$

where, EI indicates the land use mix entropy index within buffer i (SA1s), Pj represents the proportion of a type of land use j and J is the number of land use categories. Six different land-use categories, including residential, commercial, industrial, transport and infrastructure, community services and sport and recreation centres, were chosen to calculate EI. These categories were defined based on the ten main use categories defined by the Australian Valuation Property Classification Codes (AVPCC) (Morse-McNabb, 2011). Table 4 shows the list of independent variables used for analyses and their description.

Table 4. Independent variables and their description

Variables	Description
Age	Age of respondent
Sex	Gender
LNC	Driver licence
WT	Type of work
HHS	Usual number of residents in household
HHSR	Demographic structure of household
Car	Number of vehicles in household
DT	Week days/Weekend days
DWT	Type of dwelling
YL	No. of years lived at the same address
PTAI	Public Transport Accessibility Index
SI	Supply Index
PTAL	Public Transport Accessibility Level
RM	Roadway Measure
EI	Land Use mix entropy index

Note: HHSR was converted to five dummy variables: sole person, couple no children, couple with children, one parent and other; WT was converted into three dummy variables fixed hours, flexible Hours, rostered Shifts, work from Home and not working; DWT was defined as a dummy variable with four categories of separate house, terrace/townhouse, flat/apartments, DT, sex and LNC were defined as binary variables.

Table 5 suggests the descriptive statistics for the continuous variables used in the NBR models. These statistics were calculated for 18701 public transport trips. In terms of socio-demographic characteristics, the respondents were almost 38 years old on average. The average HHS shows that respondents were almost all from households with the usual number of three residents. Households lived in their address for an average of approximately ten years.

In order to examine the applicability of the PTAI compared to existing approaches, three NBR models were estimated. All the variables were considered constant in the models with the exception of the public transport accessibility measures. The PTAI and other variables were employed to run model M1, and SI was used in M2 and the PTAL in M3 (see Table 6). The coefficient values for public transport measurements are different in the models, and the PTAI in M1 has the highest value. The Incident Rate Ratio (IRR) was also calculated for the confidence level. IRR in the models describes the percentage change in the incident rate of the response variable for every unit increase in the corresponding explanatory variable (Hilbe, 2008). This shows that by one unit increase in PTAI, public transport trips

Table 5. Descriptive statistics for continuous variables

Variable	Mean	S.D.	Min	Max
PT Trips	32.09	22.08	1.00	110.00
Age	37.55	19.76	0.00	96.00
HHS	3.25	1.35	1.00	6.00
Car	1.90	0.95	0.00	4.00
YL	9.50	10.64	0.00	75.00
RM (m)	1.36	0.79	0.00	5.57
EI	0.42	0.15	0.00	0.87
PTAI	33.26	360.30	0.00	7,235.57
SI	17,191.58	17,132.71	0.00	222,037.92
PTAL	16.40	174.80	0.00	3,482.64

n=18,701 trips

increase by about 11% (IRR = 1.11). This percentage is 9% and 6% for SI and PTAL, respectively.

On the other hand, based on the goodness of fit criteria M1 has the lowest Akaike information criterion (AIC), which is a measure of the relative quality of statistical models for a given set of data. Given a series of models for the data, the AIC estimates the quality of each model relative to that of each of the other models. Hence, the AIC provides a means for model selection (Boisbunon et al., 2014, Hu, 2007, Aho et al., 2014). Regarding other factors, as shown in Table 6, living as a sole person or being a single parent are negatively associated with public transport trips. Also, by increasing number of cars in the households' public transport trips decrease. Built environment features also have a significant impact on the number of public transport trips. EI and public transport access measures are positively, and RM negatively associated with PT trips. For instance, there is an 18% increase of PT trips when EI increases by one unit. In contrast, while the RM decreases by one unit, PT trips decrease by about 13%.

1.4 DISCUSSIONS

This chapter has focused on presenting an approach to measure public transport accessibility in metropolitan Melbourne. This approach introduced as a combined measure of public transport services frequency and population

Table 6. Outputs of the NBR models for public transport trips.

Parameters	M_1			M_2			M_3		
	Coef.	Std. Err.	IRR	Coef.	Std. Err.	IRR	Coef.	Std. Err.	IRR
Cons.	3.603***	0.269	36.714	3.660***	0.269	38.859	3.845***	0.268	46.762
Age	0.000	0.001	1.000	0.000	0.001	1.000	0.000	0.001	1.000
Sex (Male)	-0.011	0.015	0.989	-0.016	0.015	0.984	-0.016	0.015	0.984
LNC (Yes)	-0.006	0.020	0.994	-0.003	0.020	0.997	0.004	0.020	1.004
HHS	0.038***	0.009	1.039	0.042***	0.009	1.043	0.040***	0.009	1.041
HHSR									
Sole Person	-0.025	0.037	0.975	-0.024	0.037	0.976	-0.048	0.037	0.953
Couple no Children	0.047*	0.029	1.048	0.037	0.029	1.038	0.021	0.029	1.021
Couple with Children	0.046*	0.026	1.047	0.018	0.026	1.018	0.011	0.026	1.011
Single Parent	-0.069**	0.034	0.934	-0.095**	0.034	0.910	-0.121***	0.034	0.886
Car	-0.056***	0.009	0.945	-0.060***	0.009	0.941	-0.075***	0.009	0.928
WT									
Fixed Hours	0.022	0.020	1.022	0.038*	0.020	1.039	0.034*	0.020	1.035
Flexible Hours	0.059**	0.025	1.061	0.057**	0.025	1.058	0.072**	0.025	1.074
Rostered Shifts	0.066**	0.029	1.069	0.080**	0.029	1.083	0.067**	0.029	1.070
Work from Home	0.142	0.072	1.153	0.115*	0.072	1.122	0.184**	0.072	1.202
DWT									
Separate House	-0.831***	0.261	0.435	-0.729**	0.261	0.482	-0.912***	0.261	0.402
Terrace/Townhouse	-0.666**	0.262	0.514	-0.568**	0.262	0.566	-0.705**	0.262	0.494
Flat/Apartments	-0.731**	0.261	0.482	-0.622**	0.262	0.537	-0.767**	0.261	0.464
Day Type (Weekdays)	-0.110***	0.023	0.896	-0.123***	0.023	0.884	-0.117***	0.023	0.890
YL	0.002*	0.001	1.002	0.002**	0.001	1.002	0.003**	0.001	1.003
EI	0.166***	0.013	1.181	0.152***	0.014	1.164	0.196***	0.013	1.216
RM	-0.135***	0.013	0.874	-0.126***	0.013	0.882	-0.121***	0.013	0.886
PTAI	0.105***	0.005	1.111						
SI				0.085***	0.005	1.089			
PTAL							0.055***	0.005	1.056

Note: (1) number of public transport trips is defined as a count dependent variable.
(2) The NBR dispersion parameter was estimated by maximum likelihood.
(3) Significance codes: p<0.001 '***' 0.01 '**'.
(4) Overall goodness-of-fit:
M1: Log Likelihood = -83240.497; AIC = 166524.993; BIC = 166697.392,
M2: Log Likelihood = -83282.169; AIC = 166608.338; BIC = 166780.738,
M3: Log Likelihood = -83369.955; AIC = 166783.909; BIC = 166956.308.

density. The PTAI was calculated for Melbourne's 9510 statistical areas. Thereafter, the index was assessed to see whether there is any significant difference between the level of accessibility and the use of public transport modes. Overall, based on the results, PTAI was found as a valid mean of measuring public transport mode use in the Melbourne region using VISTA dataset. Furthermore, the PTAI showed greater association with PT trips compared to its counterparts.

As discussed previously, approximately 30% of the Melbourne region is covered by public transport walking catchments. This includes approximately 17,800 bus stops, 1,700 tram stops and 240 train stations, with an average frequency of 2.2, 2.9 and 4.5 (per hour), respectively. Although public buses have the highest catchment coverage and frequency during peak hours, they are used less than trains (by 8.3%) and trams (by about 1%). The results indicate that 0.55% of SA1s have zero accessibility to public transport. 14% and 17% of SA1s showed very poor and poor access to public transport. Areas with lower levels of access to public transport mainly relate to the outer parts of the Melbourne region; however, some areas in inner Melbourne are not excluded from poor accessibility.

PTAI was calculated using spatial data of Melbourne region. After building up the index, the VISTA was used to evaluate the proposed index. The PTAI was compared with two common approaches, SI and PTAL, measuring public transport accessibility. For this purpose, those indexes were built for Melbourne's SA1s. Results show consistency with these indices. Although the approaches used in these studies are different, there are clear similarities between the results. There were also similarities with the results of other research which calculate public transport accessibility levels (Kerrigan and Bull, 1992, Wu and Hine, 2003). Thereafter, PTAI, SI and PTAL were employed in three separate NBR models to see how these indexes can affect PT trips. IRRs from the models indicated that PTAI had greater impact on PT trips. Moreover, model selection criteria, AIC and BIC, showed that the model including PTAI had better performance compared to the ones including SI and PTAL.

Overall, Poor access to public transport can prevent access to different facilities and social activities. Hence, from a social planning perspective, accessibility can be considered as a measure of locational disadvantage. As Lucas (2012) argued that there are inter-relationships between transport shortcomings and key areas of social disadvantage such as unemployment, health inequality and poor education. For this reason, in some transport studies, weighted socioeconomic factors are combined to calculate levels of accessibility to public transport (Currie and Stanley, 2007, Hurni, 2005). However, employing a weighted accessibility index in regression models along with socioeconomic explanatory variables may duplicate the effects of social factors and bias the results. In contrast, PTAI which reflects the service frequency of transport modes and distribution of the population can be employed as a spatial factor along with social variables in travel behaviour modelling.

1.5 CONCLUSION AND FUTURE RESEARCH DIRECTIONS

This study proposed a new approach measuring access to public transport utilized GIS techniques. The PTAI provides a practical means of measuring levels of accessibility within metropolitan areas. The approach has been evaluated and compared with previous methods. In general, the quantitative approaches developed in this study, are straightforward to apply and can be employed for any number of public transport modes in any other geographical scale. The new method is designed to be calculated with available census data and transport modelling tools. Furthermore, the analysis provides reliable and defendable results, and accessibility could be measured for 99.4% of statistical areas. In the present study, used mesh blocks as smallest geographical units available for Melbourne region to calculate population density. This greatly improved the accuracy of the results. If SA1s were used for estimating buffer populations, the results could be miscalculated by the variation of -187 to 360 persons. In other words, buffer populations would be under-estimated or over-valued by 12 persons per SA1. Nonetheless, the accuracy can be enhanced more by using parcel-based data.

A weakness of this approach is that the study does not consider the effects of temporal disparity (Neutens et al., 2012, Chen et al., 2014, Kwan, 2013) in public transport accessibility. Future studies may take into account this point when measuring accessibility. Furthermore, the PTAI does not consider connectivity between public modes, which can influence accessibility, particularly in areas of low accessibility.

REFERENCES

Altman, J., & Hinkson, M. (2007). Mobility and Modernity in Arnhem Land The Social Universe of Kuninjku Trucks. *Journal of Material Culture*, *12*(2), 181–203. doi:10.1177/1359183507078122

Brewer, C. A., & Pickle, L. (2002). Evaluation of methods for classifying epidemiological data on choropleth maps in series. *Annals of the Association of American Geographers*, *92*(4), 662–681. doi:10.1111/1467-8306.00310

Ceder, A., Net, Y., & Coriat, C. (2009). Measuring public transport connectivity performance applied in Auckland, New Zealand. *Transportation Research Record: Journal of the Transportation Research Board*, *2111*(1), 139–147. doi:10.3141/2111-16

Chen, S., Claramunt, C., & Ray, C. (2014). A spatio-temporal modelling approach for the study of the connectivity and accessibility of the Guangzhou metropolitan network. *Journal of Transport Geography*, *36*, 12–23. doi:10.1016/j.jtrangeo.2014.02.006

Cheng, J., & Bertolini, L. (2013). Measuring urban job accessibility with distance decay, competition and diversity. *Journal of Transport Geography*, *30*, 100–109. doi:10.1016/j.jtrangeo.2013.03.005

Cheng, Y.-H., & Chen, S.-Y. (2015). Perceived accessibility, mobility, and connectivity of public transportation systems. *Transportation Research Part A, Policy and Practice*, *77*, 386–403. doi:10.1016/j.tra.2015.05.003

Currie, G. (2010). Quantifying spatial gaps in public transport supply based on social needs. *Journal of Transport Geography*, *18*(1), 31–41. doi:10.1016/j.jtrangeo.2008.12.002

Currie, G., & Stanley, J. (2007). *No way to go: Transport and social disadvantage in Australian communities*. Academic Press.

Delmelle, E. C., & Casas, I. (2012). Evaluating the spatial equity of bus rapid transit-based accessibility patterns in a developing country: The case of Cali, Colombia. *Transport Policy*, *20*, 36–46. doi:10.1016/j.tranpol.2011.12.001

Department of Premier and Cabinet. (2015). *Victorian Government Data Directory*. Retrieved from www.data.vic.gov.au

EDJTR. (2009). PALI Transport, *Economic Development, Jobs, Transport and Resources*. Victorian Integrated Survey of Travel and Activity (VISTA).

Elias, W., & Shiftan, Y. (2012). The influence of individual's risk perception and attitudes on travel behavior. *Transportation Research Part A, Policy and Practice*, *46*(8), 1241–1251. doi:10.1016/j.tra.2012.05.013

Fransen, K., Neutens, T., Farber, S., De Maeyer, P., Deruyter, G., & Witlox, F. (2015). Identifying public transport gaps using time-dependent accessibility levels. *Journal of Transport Geography*, *48*, 176–187. doi:10.1016/j.jtrangeo.2015.09.008

Goodman, L. A., & Kruskal, W. H. (1954). Measures of association for cross classifications. *Journal of the American Statistical Association*, *49*, 732–764.

Hanneman, R. A., Kposowa, A. J., & Riddle, M. D. (2012). *Basic Statistics for Social Research*. John Wiley & Sons.

Hurni, A. (2005). *Transport and social exclusion in Western Sydney*. Australian Transport Research Forum (ATRF), 28th, 2005, Sydney, New South Wales, Austalia.

Ian, H. (2010). *An introduction to geographical information systems*. Pearson Education India.

Johnson, V., Currie, G., & Stanley, J. (2011). Exploring transport to arts and cultural activities as a facilitator of social inclusion. *Transport Policy*, *18*(1), 68–75. doi:10.1016/j.tranpol.2010.06.001

Kerrigan, M., & Bull, D. (1992). Measuring accessibility: A public transport accessibility index. *Environmental Issues. Proceeding of 20th Annual Seminar B Held at the PTRC European Transport, 354*.

Kwan, M.-P. (1998). Space-time and integral measures of individual accessibility: A comparative analysis using a point-based framework. *Geographical Analysis*, *30*(3), 191–216. doi:10.1111/j.1538-4632.1998.tb00396.x

Kwan, M. P. (1999). Gender and individual access to urban opportunities: A study using space–time measures. *The Professional Geographer*, *51*(2), 210–227. doi:10.1111/0033-0124.00158

Kwan, M.-P. (2013). Beyond Space (As We Knew It): Toward Temporally Integrated Geographies of Segregation, Health, and Accessibility: Space–Time Integration in Geography and GIScience. *Annals of the Association of American Geographers*, *103*(5), 1078–1086. doi:10.1080/00045608.2013.792177

Kwan, M.-P. (2015). *Beyond Space (As We Knew It): Toward Temporally Integrated Geographies of Segregation, Health, and Accessibility. In Space-Time Integration in Geography and GIScience*. Springer.

Lei, T., & Church, R. (2010). Mapping transit-based access: Integrating GIS, routes and schedules. *International Journal of Geographical Information Science*, *24*(2), 283–304. doi:10.1080/13658810902835404

London, T. F. (2010). *Measuring Public Transport Accessibility Levels*. Academic Press.

Lucas, K. (2011). Making the connections between transport disadvantage and the social exclusion of low income populations in the Tshwane Region of South Africa. *Journal of Transport Geography*, *19*(6), 1320–1334. doi:10.1016/j.jtrangeo.2011.02.007

Lucas, K. (2012). Transport and social exclusion: Where are we now? *Transport Policy*, *20*, 105–113. doi:10.1016/j.tranpol.2012.01.013

Mavoa, S., Witten, K., Mccreanor, T., & O'Sullivan, D. (2012). GIS based destination accessibility via public transit and walking in Auckland, New Zealand. *Journal of Transport Geography*, *20*(1), 15–22. doi:10.1016/j.jtrangeo.2011.10.001

Mazloumi, E., Moridpour, S., Currie, G., & Rose, G. (2011). Exploring the value of traffic flow data in bus travel time prediction. *Journal of Transportation Engineering*, *138*(4), 436–446. doi:10.1061/(ASCE)TE.1943-5436.0000329

Mitra, R., & Buliung, R. N. (2012). Built environment correlates of active school transportation: Neighborhood and the modifiable areal unit problem. *Journal of Transport Geography*, *20*(1), 51–61. doi:10.1016/j.jtrangeo.2011.07.009

Neutens, T., Delafontaine, M., Scott, D. M., & De Maeyer, P. (2012). An analysis of day-to-day variations in individual space–time accessibility. *Journal of Transport Geography*, *23*, 81–91. doi:10.1016/j.jtrangeo.2012.04.001

Priya, T., & Uteng, A. (2009). Dynamics of transport and social exclusion: Effects of expensive driver's license. *Transport Policy*, *16*(3), 130–139. doi:10.1016/j.tranpol.2009.02.005

Saghapour, T., Moridpour, S., & Thompson, R. G. (2016). Public Transport Accessibility in Metropolitan Areas: A New Approach Incorporating Population Density. *Journal of Transport Geography*, *54*, 273-285.

Saghapour, T., Moridpour, S., & Thompson, R. G. (2017). Modeling Access to Public Transport in Metropolitan Areas. *Journal of Advanced Transportation*, *50*(8), 1785-1801.

Taniguchi, E., Thompson, R. G. & Yamada, T. (2013). Concepts and Visions for Urban Transport and Logistics Relating to Human Security. *Urban Transportation and Logistics: Health, Safety, and Security Concerns*, 1.

Wang, C.-H., & Chen, N. (2015). A GIS-based spatial statistical approach to modeling job accessibility by transportation mode: Case study of Columbus, Ohio. *Journal of Transport Geography*, *45*, 1–11. doi:10.1016/j.jtrangeo.2015.03.015

Wong, D. (2009). The modifiable areal unit problem (MAUP). The SAGE handbook of spatial analysis, 105-123.

Wu, B. M., & Hine, J. P. (2003). A PTAL approach to measuring changes in bus service accessibility. *Transport Policy*, *10*(4), 307–320. doi:10.1016/S0967-070X(03)00053-2

ENDNOTE

1 According to the Australian Bureau of Statistics (ABS), the ABS structure of Melbourne region contains 53074 Mesh Blocks, 9510 Statistical Areas Level 1 (SA1s), 277 Statistical Areas Level 2 (SA2s), 42 Statistical Areas Level 3 (SA3s) and 12 Statistical Areas Level 4 (SA4s).

Chapter 2
Bikeability in Metropolitan Areas

ABSTRACT

There have been several techniques for measuring bikeability; however, limited comprehensive research has been conducted focusing on travel distance as an important barrier for cyclists. Furthermore, existing measurements are mainly restricted by the availability of travel behaviour data. In this chapter, a new index for measuring bikeability in metropolitan areas is presented. The Cycling Accessibility Index (CAI) has been developed for computing cycling accessibility within Melbourne metropolitan, Australia. The CAI is defined consistent with gravity-based measures of accessibility. This index measures cycling accessibility levels considering mixed use developments as well as travel distance between origins and destinations. The Victorian Integrated Survey of Travel and Activity (VISTA) dataset was used to assess the proposed index and investigate the association between cycling accessibility levels and number of bicycle trips in local areas. Key findings indicate that there is a significant positive association between bicycle trips and the CAI.

2.1 INTRODUCTION

Promoting non-motorised accessibility has recently become an important objective for urban and transport planners (Iacono et al., 2010, Vale, 2013). Previous research on bicycle accessibility to destinations indicates that people commonly exclude potential destinations because of distance and travel time.

DOI: 10.4018/978-1-5225-7943-4.ch002

Most studies consider travel distance between the origins and destinations as travel impedance. In these studies, accessibility reflects the attractiveness of facilities weighted by the travel time needed to reach those destinations (Sun et al., 2012, Hull et al., 2012, Silva and Pinho, 2010). However, travel distance has also been considered as travel impedance in some studies (Iacono et al., 2010, Lowry et al., 2012, Vasconcelos and Farias, 2012). Lowry et al. (2012) introduced a bikeability index which focused on bicycle trips. This study assessed the bikeability of the entire road network in terms of access to important destinations.

One practical reason of considering gravity-based or location-based accessibility measures for non-motorised modes of transport is their potential compatibility with regional travel forecasting models. Hence, they can easily extract travel times from one zone to another based on coded networks. In addition, a number of potential opportunities are available at the zone level (Iacono et al., 2010). However, one of the limitations of the use of these measures for non-motorised modes relates to the use of non-motorised modes in travel demand models. With respect to travel time, motorised modes are more sensitive to travel times and levels of network congestion than non-motorised modes of transport. Furthermore, non-motorised route choice tends to include factors that may be more qualitative, experiential or difficult to measure/quantify (Iacono et al., 2010, Tilahun et al., 2007, Hunt and Abraham, 2007).

Another limitation of existing approaches that measure cycling accessibility is that they are highly dependent on travel diary data. In addition, methods that have been applied to measuring cycling accessibility have not focused on the cycling availability of destinations in terms of service coverage areas. Some of the measures have focused on determining the level of service in terms of network infrastructure, such as the Bicycle Compatibility Index (BCI) or the Bicycle Level of Service (BLOS) for a bicycle network (Harkey et al., 1998a, Harkey et al., 1998b, Landis et al., 1997, Landis et al., 2003). These studies measure the performance of a bicycle network using various geometric measures, such as the width of the bicycle routes, pavement, route types, and connectivity. However, there are other methods that consider bikeability in terms of how accessible different destinations are for bicycles as a transport mode. Such methods measure the potential for cycling using travel behaviour data (Rybarczyk and Gallagher, 2014, Wahlgren and Schantz, 2012, Milakis et al., 2015, Espada and Luk, 2011).

As mentioned above, many of destinations may be eliminated from a user's choice because of the distance. A question that arises here is that how people define an acceptable distance for using bicycle use and how this threshold affects the levels of bikeability to destinations (Milakis et al., 2015, Rahul and Verma, 2014). Several factors, such as gender (Akar et al., 2013, Bonham and Wilson, 2012), exogenous restrictions (such as danger, vandalism and facilities) (Fernández-Heredia et al., 2014), safety (Mesbah et al., 2012), stress in terms of traffic volume and speed (Sorton and Walsh, 1994, Mekuria et al., 2012, Lowry et al., 2016), the relationship between commuting time and work duration (Schwanen and Dijst, 2002) and the time needed to spend on other activities (Hupkes, 1982), have been identified as the main factors influencing acceptable cycling travel time (Milakis et al., 2015). Existing studies on cycling accessibility mainly focus on access to some specific destinations such as employment as an important factor in forming urban structures. However, there are limited research which considers access to other destinations, such as retail, recreation and education, can also influence travel behaviour (Daly, 1997, Iacono et al., 2010).

Although non-motorised accessibility to various destinations has recently emerged as an important topic planning (Iacono et al., 2010, Krizek, 2005), most measures introduced are not comprehensive (Iacono et al., 2010). The main limitation of existing measurements is that they are highly dependent on travel data by non-motorised modes. In contrast, travel data for non-motorised transportation are limited, and in most cases, they are questionnaire-based and may not be reliable. The provision of consistent and robust metrics for accessibility offers a defendable foundation for sustainability policy regarding travel and the built environment. In this regard, introducing accurate accessibility measures for walking or cycling should assist transport planners in making more rational decisions in infrastructure provision for non-motorised transportation (Devkota et al., 2012, Iacono et al., 2010, Levine, 2010). This paper introduces an index that measures the level of cycling accessibility within geographical areas. It demonstrates how cycling access to different destinations can be reliably measured. The cycling accessibility index has been developed by using spatial dataset of metropolitan Melbourne, Australia.

Section 2.2 describes the methodology developed. The methodology section is followed by analysis and results of application of the CAI in the Melbourne metropolitan region in Section 2.3. Section 2.4 discusses the results and section 2.5 summarizes the findings and outlines avenues for future research.

2.2 METHODS

This chapter proposes an index for measuring cycling accessibility levels in Melbourne's 9510 Statistical Areas level 1 (SA1s). In measuring cycling accessibility, two factors, distance or travel time between origins and destinations and the cycling catchments of destinations are considered. Cycling catchments are calculated based on the service area of destinations and the travel distance, which is considered as the distance between origins and destinations. Network models are applied to identify acceptable cycling catchments as well as an origin-destination (O-D) cost matrix of origins and destinations using a geographical information system (GIS). The calculation procedure is fully explained in the approach section. The databases, study area and conceptual framework are presented in the following sections which describe the procedure for calculating the index.

2.2.1 Datasets

As explained in previous sections, the aim of this part of the study was to measure cycling accessibility within Melbourne's 9510 Statistical Areas Level 1 (SA1s). For this purpose, several datasets were adopted, which are described as follows.

Geographical Areas

As explained in previous chapter a database of mesh blocks from the 2011 Census for the Melbourne Region is available from the Australian Bureau of Statistics (ABS, 2011). Analyses for calculating the bikeability within the Melbourne was based on the SA1 level.

Point of Interests (POIs)

A database of points of interest (POIs) was obtained from PSMA Australia (PSMA, 2011b). POIs include urban centres, significant buildings, landmarks; public spaces, community facilities and indigenous locations, and those for Melbourne include 15,588 points. These POIs are considered as into four groups of activities including education centres, health and care facilities, retail and recreation centres, community services.

Principal Bicycle Network

The Bicycle Network dataset was obtained from the Victorian Government open data sources (Data.Vic, 2015). This dataset was produced by the Roads Corporation of Victoria (VicRoads) in 2015, however, the last verification date was in 2011. The dataset contains information on the 4,139 bicycle path segments with an average length of 1 km. Figure 1 presents the bicycle network in the Melbourne metropolitan area.

VISTA Dataset

The VISTA dataset (Transport, 2009) as explained in previous chapter was used for evaluating the proposed index. In this research, only trips made by bicycle were used for analysis. According to the dataset 1340 trips were reported as bike trips.

Figure 1. Principal bicycle network within the Melbourne region

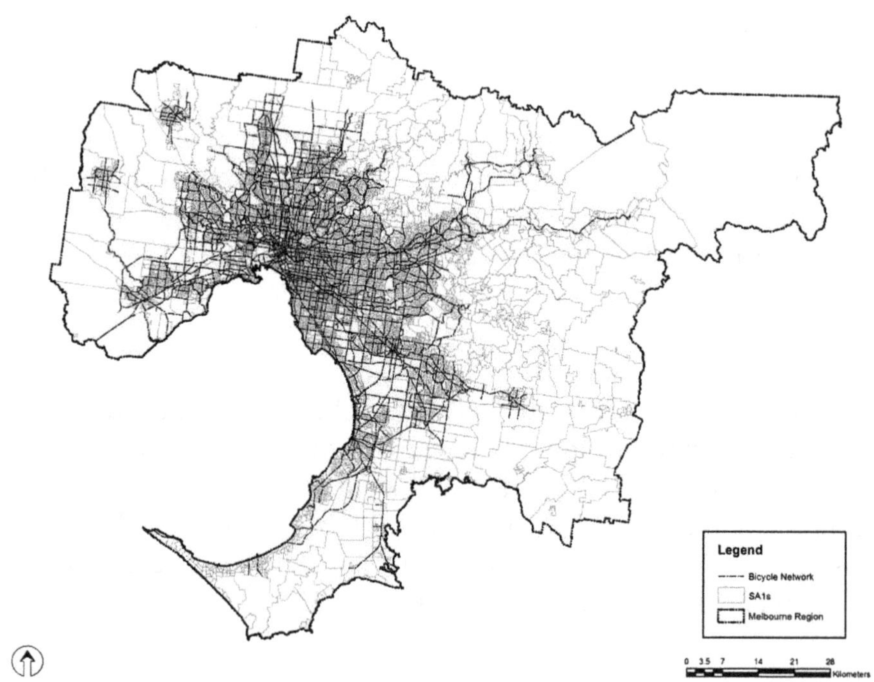

2.2.2 Calculation Framework

The CAI calculation procedure has two main steps. The first step relates to the SA1s' weighted centroids as origins, and the second step involves calculations relating to the destination groups. Figure 2. illustrates a framework of the CAI calculation procedure. A network analysis including service area and O-D cost matrix analysis, using the ArcGIS 10.2 software, are applied to calculate both the cycling catchments of the destination groups and travel distances between origins and destinations ($\bar{D}_{ij}$). Thereafter, the ratio of cycling catchment areas to SA1 areas on one side and the ratio of $\bar{D}_{ij}$ to bicycle path lengths within SA1s on the other side are used to compute the Cycling Accessibility Index.

SA1s Weighted Centroids[1]

SA1s with an average population of 414 people and average area of 1 square km are built up from or, are approximated by, the mesh blocks and each SA1 contains five mesh blocks on average. The mesh block with the highest

Figure 2. A framework of CAI calculation procedure

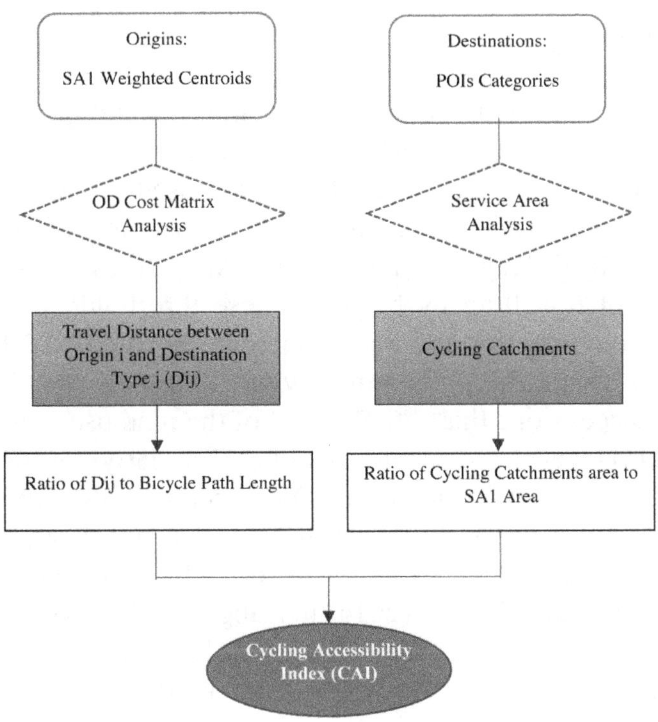

Figure 3. Weighted centroids for SA1s

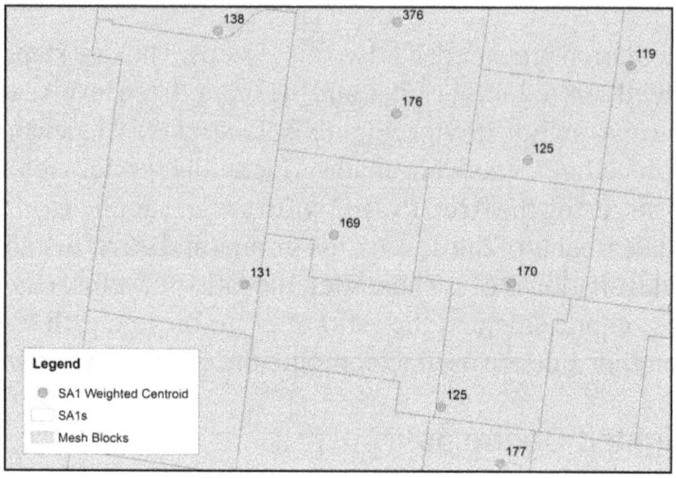

population within the corresponding SA1 is defined as a weighted centroid of the SA1 and it is considered as an origin. Figure 3 illustrates mesh blocks, SA1s and weighted centroids. As the figure indicates, the centroid for the selected SA1 is placed on the mesh block with the highest population.

Destinations

As explained in previous section, destinations were grouped into four categories. Service area and OD-cost matrix analysis was undertaken for each set of destinations separately. This study used the Median Desirable Travel Distances (MDTD) defined by the Association of Australian and New Zealand Road Transport and Traffic Authorities (Austroads, 2011) as thresholds of travelling by bicycle. These thresholds were defined as 4km for education centres and health and care facilities, 2.5 for retail and recreation centres and 5.2 for community services. According to Austraods, a MDTD distance is one that satisfies half of the road users. In service area analysis, MDTD were used as cut-off values. Similarly, for travel distances between origins and destinations, an OD-cost matrix was applied separately for each type of destinations. MDTDs used in the current study were found to be more consistent with the findings reported by McDonald (2007) in the United States, Rahul & Verma (2014) in Bangalore, Milakis et al. (2015) in Berkeley, Iacono et al. in the United States (2010) and McNeil (2011) in the United States. The average speed of 15 to 16 km/h has been considered in previous research for cyclists (Espada et al., 2015, Espada and Luk, 2011, Prud'homme and Bocarejo, 2005, Paris, 2010). This study uses the speed of 16 km/h, adopted from the Austroads network operation planning framework (Espada and Luk, 2011).

2.2.3 Approach

As explained in the previous section, cycling catchments and travel distances between origins and destinations were calculated considering the MDTDs. Travel distances between origins and destinations are required to identify how far destinations are located from the origins (weighted centroid of SA1s) (Saghapour, Moridpour, & Thompson, 2017).

Cycling catchments were calculated for each destination type based on the MDTDs. For this purpose, the service area of network analyst tools in ArcGIS 10.2 was used. Cycling catchments were calculated for each category of destinations, considering the related MDTD as the cut-off value. In the next step, the ratio of cycling catchments in each group to the area of the corresponding SA1were calculated. Computed ratios were then summed to represent Area Ratio (AR_i) as formulated in Equation 1.

$$AR_i = \sum_{j=1}^{4} N_{ij} * \left(\frac{Area_{CC_{ij}}}{Area_i} \right) \quad (1)$$

where, AR_i is the ratio of cycling catchment areas to the area of the $SA1_i$, N_{ij} is the total number of cycling catchment areas for destination type j in $SA1_i$, $Area_{ccij}$ denotes the cycling catchment area for destination type j in $SA1_i$, and $Area_i$ indicates the area of the $SA1_i$. The Area Ratio (AR) defined in Equation 2 measures both the diversity and intensity of land uses. Given that *N* in Equation 2 denotes the number of activities available for cyclists within cycling catchments; *AR* reflects the intensity of different land uses. Hence, the more activities available, the higher the value of *AR* calculated. On the other hand, the total value of the *AR* reflects the diversity of land uses, because it is computed by summing the *AR* values of all destination categories. In other words, for a given *SA1,* if the number of destinations available within an acceptable distance is doubled, the total value of *AR* is also doubled for a constant value of $Area_{ccj}/Area_i$.

For each SA1, the CAI was calculated using the formula shown in equation 2. This index is a combined measure of AR_i and exponential function of X_{ij} given as:

$$CAI_i = AR_i + \sum_{j=1}^{4} e^{-X_{ij}} \left(X_{ij} = \frac{\bar{D}_{ij}}{Bl_i}, N_{ij} \neq 0 \right) \tag{2}$$

where CAI_i is the Cycling Accessibility Index for each $SA1$, AR_i is the ratio of the combined cycling catchment areas in $SA1_i$ to the total area of $SA1_i$, X_{ij} is the travel impedance which is the ratio of average travel distance between origin i and destination type j to the total bicycle length in the corresponded SA1. For the areas with no bicycle network, the CAI is equal to AR_i. The logic behind this is that cyclists may share the road with other modes within those areas. For areas with no destinations within a MDTD, the value is zero. In other words, if $N_{ij} = 0$, then X_{ij} and $\bar{D}_{ij}$ are undefined. More details and illustrations of the calculation of the CAI is provided in the study by Saghapour et al. (2016a).

2.3 RESULTS

Table 1 presents the ranges and categories of the CAI. The index is grouped into four main categories: poor, moderate, good, and excellent, and a zero group. The classification method used for the CAI categories is based on quintiles (Espada and Luk, 2011, TfL, 2010) since they are one of the best methods for simplifying comparisons as well as aiding general map-reading (Brewer and Pickle, 2002). Zero accessibility is estimated for 86,929 residents or 2.6% of the total population. Poor accessible areas are mostly located in outer Melbourne. Overall, around 50% of the total population has zero to moderate cycling accessibility.

Figure 4 shows the distribution of CAI within the Melbourne region. As explained above, the CAI is categorized into four bands. However, the first and last categories are further sub-divided into sub-levels to provide better clarity. The first category represents poor accessibility while the last category corresponds to an excellent level of bikeability. High levels of accessibility from good to excellent are mostly concentrated in the inner parts of the Melbourne.

Table 1. CAI ranges and categories

CAI Categories	CAI Ranges	SA1s		Population	
		No	%	No	%
NA/Zero	0	246	2.6	86,929	2.2
Poor	< 0.5	2,013	21.2	819,933	20.8
Moderate	0.5 - 2	2,560	26.9	1,072,778	27.2
Good	2 – 4.5	2,452	25.8	1,023,326	26.0
Excellent	> 4.5	2,239	23.5	938,245	23.8
Total	NA	9,510	100.0	3,941,211	100.0

Evaluation of the CAI

As mentioned previously, VISTA dataset records travel in the form of trip stages, where a "trip stage" is a segment of travel with a single purpose and mode. Hence, the dataset contains details of 93,902 trips stages made by 22,184 individuals in the MSD, however, only 1,340 or 1.4% of total trips were made by bicycle. These numbers of bicycle trips belong to 320 numbers of SA1s. Considering the total number of bicycle trips (1,340), the average number of bicycle trips in each SA1 is about 4. Followings present the results of analysis conducted to assess the CAI practicality, using a real travel data. Table 2 shows number and percentage of trips made by different modes in the Melbourne region.

Outlier Detection for CAI Values

Prior to running the analysis for evaluating the CAI, the calculated values of the index were tested for outliers. Median and mean absolute values were selected to identify outliers. Figure 5 presents the box and whisker plots for the CAI values.

Outlier detection results as presented in the Table 3, 71 or 5% of values were designated as outliers. Hence, after removing the outliers, the number of observations used for analysis was 1269.

Tests of Association

CAI values were joined to the travel data suing the SA1 unique code. Thereafter, Chi-Square tests were applied to examine whether there was

Figure 4. CAI within the Melbourne region

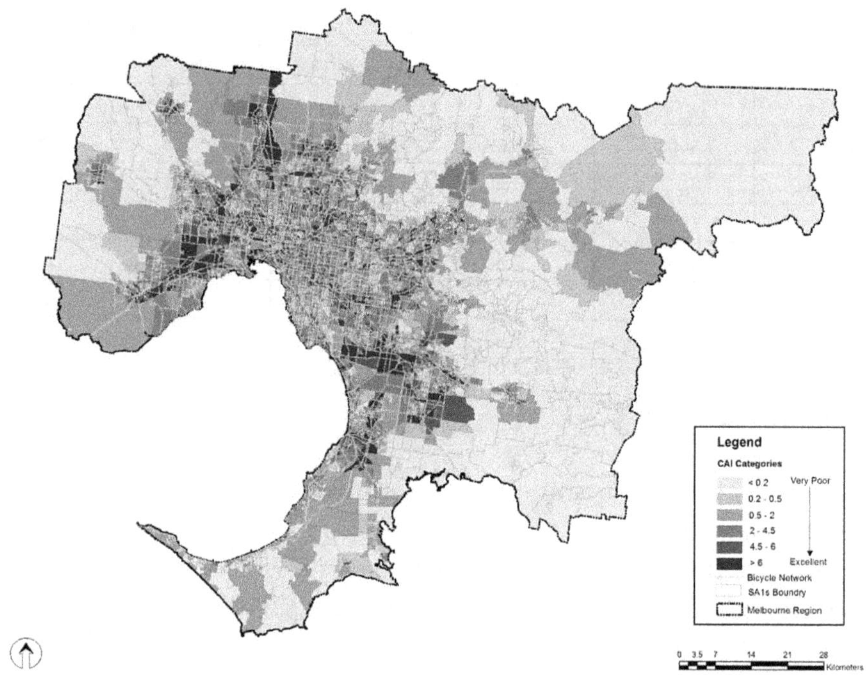

Table 2. Number and percentage of trips made by different modes in the Melbourne region

Transport Modes	Frequency	Percentage
Motorised Vehicle	64236	68.41
Walking	9,625	10.25
Bicycle	1,340	1.43
Public Transport	18324	19.51
Other	377	0.40
Total	93,902	100.0

any association between the proposed accessibility index and the number of bicycle trips. Table 4 presents the results of associations' tests between CAI categories and bicycle trips. According to the content of the table, CAI and number of bicycle trips in SA1s have a significant degree of association ($\chi^2 = 601.349, p < .001$).

Figure 5. Box and whisker plots for CAI values

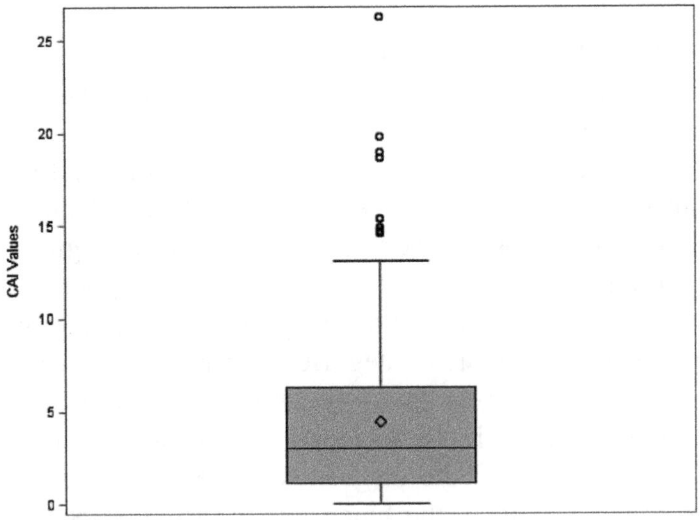

Table 3. Basic statistics and outlier analysis for CAI

Parameters	Values
Mean	4.44
Median	2.97
Std. Deviation	4.43
Number of Outliers	71
Proportion of Outliers	0.05

Sample Size: 1340

Table 4. Chi-square test for bicycle trips and CAI categories

Statistic	DF	Value	Prob.
Chi-Square	51	601.349	<.0001
Likelihood Ratio Chi-Square	51	609.548	<.0001
Mantel-Haenszel Chi-Square	1	36.585	<.0001
Phi Coefficient		0.688	
Contingency Coefficient		0.567	
Cramer's V		0.397	

Sample Size: 1269

2.4 DISCUSSIONS AND CONCLUSION

This part of the study proposed an approach for measuring bikeability within the Melbourne's 9,510 SA1s. The CAI objectively measure bikeability utilizing GIS techniques, including an OD cost matrix and service area analysis. The index presented was a combined measure of the cycling catchment areas an impedance function of travel distance between the defined origins and destinations. Destinations were categorized into four groups including health and care facilities, education centres, community services and retail and recreation centres. Service areas analyses were applied for calculating the cycling catchments named categories, separately. Service areas were then divided by the corresponding SA1's area to obtain the area ratios. For the second component of the CAI the travel impedance were computed as the travel distance between the origins, weighted centroids of SA1s, and destinations. To assure about the functionality of the proposed index, CAI was evaluated using real travel data. For this purpose, Chi-Square tests were run using CAI categories and bicycle trips. As revealed from the results there was a statistically significant association between the cycling accessibility levels and bicycle trips within statistical areas. Thus, the CAI was evaluated as a valid means of measuring bikeability in the Melbourne region based on the VISTA dataset.

Key findings indicate that 2.6% of SA1s or 2.2% of residents have no cycling accessibility and about 50% of areas have poor to moderate accessibility. Areas with zero to moderate levels of cycling accessibility were mostly belonging to the outer Melbourne; nevertheless, the inner suburbs were not excluded from low cycling accessibility. In other words, it can be concluded that the Melbourne CBD is more accessible by bicycle as a mode of transport. However, the outer suburbs with dispersed patterns have no to little coverage of bicycle networks and therefore less cycling accessibility.

In brief, from a social planning perspective, accessibility can be considered as a measure of locational status. In addition, considering sustainable transportation and the goals of promotion of active transportation, poor cycling access can deter travel to different facilities and social opportunities. This study has several strengths and limitations. A practical measure of cycling accessibility presented in this study was built up of a geometric measure of bicycle networks and travel impedance independent of travel behaviour data. In addition, this approach considered both the mixed use developments and number of activities within a specified geographical extent, while existing approaches have mostly focused on either the diversity or intensity of land uses (Devkota et al., 2012, Iacono et al., 2010). Based on existing knowledge, the diversity of land use is significantly associated with non-motorised trips (Lee et al., 2014, Cervero et al., 2009, Handy and Xing, 2011). Hence, to reflect the impacts of land use diversity, different categories of destinations were taken into account in the derivation of the CAI. On the other hand, the number of cycling catchments reflects the number of activities (intensity of land uses) in local areas, which is a significant factor in determining accessibility (Iacono et al., 2010).

2.5 FUTURE RESEARCH DIRECTIONS

The techniques presented are simple and easy to apply. The quantitative approaches developed are not limited to specific geographical area and it can be easily employed for different types and categories of destinations in other cities around the world. The new method only requires the availability of a census data and transport modelling tools. Moreover, the methods described provide defendable results, and accessibility could be measured for about 95% of statistical areas; while it measures the accessibility within a large geographical scale. Nonetheless, the CAI can be enhanced by greater detail to achieve even more accurate results. As El-Geneidy, Krizek, & Iacono (2007) and Parkin & Rotheram (2010) argued the speed of cyclists may depend some factors such as segment length, trip length, number of signalised intersections, average daily traffic, time of day and personal characteristics. However, the present study, due to the availability of data, travel distances were only calculated based on segment length.

According to Weber (2006) temporal and individual or household-level restrictions may have a significant influence on accessibility levels which a person actually experiences at a given location. In addition, considering individual-level characteristics or constraints, such as the availability of motorised/non-motorised modes, gender, household size, household structure, etc. (Fernández-Heredia et al., 2014, Damant-Sirois and El-Geneidy, 2015) would affect the relationship between accessibility and non-motorised trips. Besides, the importance of the natural environment and ecological factors, such as the weather (Motoaki and Daziano, 2015, Ortúzar et al., 2000), vegetation (Van Holle et al., 2014) and slope (Galanis et al., 2014), should not be ignored. Therefore, future research may consider the above-mentioned limitations to achieve more accurate and reliable results.

REFERENCES

Akar, G., Fischer, N., & Namgung, M. (2013). Bicycling choice and gender case study: The Ohio State University. *International Journal of Sustainable Transportation*, *7*(5), 347–365. doi:10.1080/15568318.2012.673694

Austroads. (2011). *Application of accessibility measures, AP-R397-11*. Sydney: Austroads.

Bonham, J., & Wilson, A. (2012). Bicycling and the life course: The start-stop-start experiences of women cycling. *International Journal of Sustainable Transportation*, *6*(4), 195–213. doi:10.1080/15568318.2011.585219

Brewer, C. A., & Pickle, L. (2002). Evaluation of methods for classifying epidemiological data on choropleth maps in series. *Annals of the Association of American Geographers*, *92*(4), 662–681. doi:10.1111/1467-8306.00310

Cervero, R., Sarmiento, O. L., Jacoby, E., Gomez, L. F., & Neiman, A. (2009). Influences of built environments on walking and cycling: Lessons from Bogotá. *International Journal of Sustainable Transportation*, *3*(4), 203–226. doi:10.1080/15568310802178314

Council, M. C. (2008). *Melbourne Bicycle Account: Cycling Census 2008*. City of Melbourne.

Council, M. C. (2016). *Bicycle Plan 2016-2020*. City of Melbourne.

Daly, A. (1997). Improved methods for trip generation. In *Proceedings of seminar F held at PTRC European Transport Forum*. Brunel University.

Damant-Sirois, G., & El-Geneidy, A. M. (2015). Who cycles more? Determining cycling frequency through a segmentation approach in Montreal, Canada. *Transportation Research Part A, Policy and Practice*, *77*, 113–125. doi:10.1016/j.tra.2015.03.028

Data.Vic. (2015). *Victorian Government open data sources*. Retrieved 2015, 2015, from https://www.data.vic.gov.au/

Devkota, B., Dudycha, D., & Andrey, J. (2012). Planning for non-motorised travel in rural Nepal: A role for geographic information systems. *Journal of Transport Geography*, *24*, 282–291. doi:10.1016/j.jtrangeo.2012.03.007

DfT. (n.d.). *Accessibility planning guidance*. Transport for London, Department of Transport.

El-Geneidy, A. M., Krizek, K. J., & Iacono, M. (2007). Predicting bicycle travel speeds along different facilities using GPS data: a proof of concept model. *Proceedings of the 86th Annual Meeting of the Transportation Research Board, Compendium of Papers.*

Espada, I., Bennett, P., Green, D., & Hatch, D. (2015). *Development of the accessibility-based network operations planning framework.* Academic Press.

Espada, I., & Luk, J. (2011). *Application of accessibility measures.* Academic Press.

Fernández-Heredia, Á., Monzón, A., & Jara-Díaz, S. (2014). Understanding cyclists' perceptions, keys for a successful bicycle promotion. *Transportation Research Part A, Policy and Practice, 63*, 1–11. doi:10.1016/j.tra.2014.02.013

Galanis, A., Papanikolaou, A., & Eliou, N. (2014). Bikeability Audit in Urban Road Environment: Case Study in the City of Volos, Greece. *International Journal of Operations Research and Information Systems, 5*(2), 21–39. doi:10.4018/ijoris.2014040102

Geurs, K. T., & Van Wee, B. (2004). Accessibility evaluation of land-use and transport strategies: Review and research directions. *Journal of Transport Geography, 12*(2), 127–140. doi:10.1016/j.jtrangeo.2003.10.005

Handy, S. L., & Clifton, K. J. (2001). Evaluating neighborhood accessibility: Possibilities and practicalities. *Journal of Transportation and Statistics, 4*(2/3), 67–78.

Handy, S. L., & Xing, Y. (2011). Factors correlated with bicycle commuting: A study in six small US cities. *International Journal of Sustainable Transportation, 5*(2), 91–110. doi:10.1080/15568310903514789

Hansen, W. G. (1959). How accessibility shapes land use. *Journal of the American Institute of Planners, 25*(2), 73–76. doi:10.1080/01944365908978307

Hupkes, G. (1982). The law of constant travel time and trip-rates. *Futures, 14*(1), 38–46. doi:10.1016/0016-3287(82)90070-2

Iacono, M., Krizek, K. J., & El-Geneidy, A. (2010). Measuring non-motorised accessibility: Issues, alternatives, and execution. *Journal of Transport Geography, 18*(1), 133–140. doi:10.1016/j.jtrangeo.2009.02.002

Krizek, K. J. (2005). *Perspectives on accessibility and travel.* Academic Press.

Kwan, M.-P. (1998). Space-time and integral measures of individual accessibility: A comparative analysis using a point-based framework. *Geographical Analysis*, *30*(3), 191–216. doi:10.1111/j.1538-4632.1998.tb00396.x

Kwan, M.-P. (2013). Beyond Space (As We Knew It): Toward Temporally Integrated Geographies of Segregation, Health, and Accessibility: Space–Time Integration in Geography and GIScience. *Annals of the Association of American Geographers*, *103*(5), 1078–1086. doi:10.1080/00045608.2013.792177

Kwan, M.-P., Murray, A. T., O'Kelly, M. E., & Tiefelsdorf, M. (2003). Recent advances in accessibility research: Representation, methodology and applications. *Journal of Geographical Systems*, *5*(1), 129–138. doi:10.1007101090300107

Lee, J.-S., Nam, J., & Lee, S.-S. (2014). Built environment impacts on individual mode choice: An empirical study of the Houston-Galveston metropolitan area. *International Journal of Sustainable Transportation*, *8*(6), 447–470. doi:10.1080/15568318.2012.716142

Levine, J. (2010). *Zoned out: Regulation, markets, and choices in transportation and metropolitan land use*. Routledge. doi:10.4324/9781936331215

Lowry, M. B., Furth, P., & Hadden-Loh, T. (2016). Prioritizing new bicycle facilities to improve low-stress network connectivity. *Transportation Research Part A, Policy and Practice*, *86*, 124–140. doi:10.1016/j.tra.2016.02.003

McDonald, N. C. (2007). Active transportation to school: Trends among US schoolchildren, 1969–2001. *American Journal of Preventive Medicine*, *32*(6), 509–516. doi:10.1016/j.amepre.2007.02.022 PMID:17533067

McNeil, N. (2011). Bikeability and the 20-min neighborhood: How infrastructure and destinations influence bicycle accessibility. *Transportation Research Record: Journal of the Transportation Research Board*, *2247*(1), 53–63. doi:10.3141/2247-07

Mekuria, M. C., Furth, P. G., & Nixon, H. (2012). *Low-stress bicycling and network connectivity*. Academic Press.

Mesbah, M., Thompson, R., & Moridpour, S. (2012). Bilevel optimization approach to design of network of bike lanes. *Transportation Research Record: Journal of the Transportation Research Board, 2284*(1), 21–28. doi:10.3141/2284-03

Milakis, D., Cervero, R., Van Wee, B., & Maat, K. (2015). Do people consider an acceptable travel time? Evidence from Berkeley, CA. *Journal of Transport Geography, 44*, 76–86. doi:10.1016/j.jtrangeo.2015.03.008

Motoaki, Y., & Daziano, R. A. (2015). A hybrid-choice latent-class model for the analysis of the effects of weather on cycling demand. *Transportation Research Part A, Policy and Practice, 75*, 217–230. doi:10.1016/j.tra.2015.03.017

Neutens, T., Witlox, F., Van De Weghe, N., & De Maeyer, P. (2007). Space–time opportunities for multiple agents: A constraint-based approach. *International Journal of Geographical Information Science, 21*(10), 1061–1076. doi:10.1080/13658810601169873

Ortúzar, J. D., Iacobelli, A., & Valeze, C. (2000). Estimating demand for a cycle-way network. *Transportation Research Part A, Policy and Practice, 34*(5), 353–373. doi:10.1016/S0965-8564(99)00040-3

Paris. (2010). *Moving in Paris, Mairie de Paris*. Paris.

Parkin, J., & Rotheram, J. (2010). Design speeds and acceleration characteristics of bicycle traffic for use in planning, design and appraisal. *Transport Policy, 17*(5), 335–341. doi:10.1016/j.tranpol.2010.03.001

Prud'homme, R., & Bocarejo, J. P. (2005). The London congestion charge: A tentative economic appraisal. *Transport Policy, 12*(3), 279–287. doi:10.1016/j.tranpol.2005.03.001

PSMA. (2011). *Features of Interest*. Data Product Description.

Rahul, T. M., & Verma, A. (2014). A study of acceptable trip distances using walking and cycling in Bangalore. *Journal of Transport Geography, 38*, 106–113. doi:10.1016/j.jtrangeo.2014.05.011

Schwanen, T., & Dijst, M. (2002). Travel-time ratios for visits to the workplace: The relationship between commuting time and work duration. *Transportation Research Part A, Policy and Practice, 36*(7), 573–592. doi:10.1016/S0965-8564(01)00023-4

Saghapour, T., Moridpour, S., & Thompson, R. G. (2017). Measuring Cycling Accessibility in Metropolitan Areas. *International Journal of Sustainable Transportation*, *11*(5), 381-394.

Sorton, A., & Walsh, T. (1994). Bicycle stress level as a tool to evaluate urban and suburban bicycle compatibility. *Transportation Research Record: Journal of the Transportation Research Board*, 17–17.

TfL. (2010). Measuring Public Transport Accessibility Levels. *Transport for London*. Available: http://www.webptals.org.uk/

Transport, D. o. (2009). *Victorian Integrated Survey of Travel and Activity*. Available: http://www5.transport.vic.gov.au/

Vale, D. S. (2013). Does commuting time tolerance impede sustainable urban mobility? Analysing the impacts on commuting behaviour as a result of workplace relocation to a mixed-use centre in Lisbon. *Journal of Transport Geography*, *32*, 38–48. doi:10.1016/j.jtrangeo.2013.08.003

Van Holle, V., Van Cauwenberg, J., Deforche, B., Goubert, L., Maes, L., Nasar, J., ... De Bourdeaudhuij, I. (2014). Environmental invitingness for transport-related cycling in middle-aged adults: A proof of concept study using photographs. *Transportation Research Part A, Policy and Practice*, *69*, 432–446. doi:10.1016/j.tra.2014.09.009

VicRoads. (2015). *Principal Bicycle Network (PBN)*. Vicoria, Australia: Victorian Government Data Directory.

Weber, J. (2006). Reflections on the future of accessibility. *Journal of Transport Geography*, *14*(5), 399–400. doi:10.1016/j.jtrangeo.2006.06.005

ENDNOTE

1 Population weighted centroid is a geographical term that is different from 'weighted average'. Population weighted centroid is an algorithm used in ArcGIS to obtain a summary reference point for the centre of the population.

Chapter 3
Walkability in Metropolitan Area

ABSTRACT

Promoting active trips has been considered as a key element towards achieving more sustainable transportation. Walking as a mode of transportation can contribute to more sustainable and healthy travel habits. This chapter presents a new approach for measuring walkability within Melbourne region, Australia. An integrated approach combining transport and land-use planning concepts was employed to construct the walking access index (WAI), which is a location-based measure for accessibility. The WAI along with a common existing walkability index were employed in regression models to examine how the new index performs in transport modelling. Key findings indicate that residents are more likely to have walking trips when living in a more walkable environment. Furthermore, it was found using statistical modelling that the WAI produces better results than one of the common approaches.

3.1 INTRODUCTION

A substantial body of planning studies have conducted indicating that active transportation is consistently positively associated with urban form variables, including mixed land use, street connectivity and residential density (Frank et al., 2010). On the other hand, promoting active transportation has recently attracted a considerable attention by the health practitioners (Frank et al., 2004, Ewing et al., 2003, Saelens et al., 2003). Walking is known as the most

DOI: 10.4018/978-1-5225-7943-4.ch003

common moderate-intensity activity of adults, and is found to be associated with significant health benefits (Manson et al., 1999, Hayashi et al., 1999).

Several definitions are found for "walkability" or "walkable" neighbourhoods. Bauman et al. (Bauman et al., 2012) argued that walkable neighbourhoods are designed in a way that residents can walk from home to nearby destinations. Manaugh and El-Geneidy (2011) claimed that walkability can be defined as a ''match'' between residents' desires and expectations for various types of destinations, their willingness to walk a given distance and the quality of the required path. Hence, neighbourhoods that have this match between the form of the built environment, and residents' needs will likely have higher rates of walking trips. In another study, Frank et al. (Frank et al., 2010) defined walkability as proximity from home to non-residential destinations and concluded people living in walkable neighbourhoods are less likely to be overweight or obese than people living in more suburban areas that require motorised transportation.

Improving the built environment to make it more convenient for people to be physically active, is an essential component of increasing physical activity (Dannenberg et al., 2003, Frank et al., 2003, Lavizzo-Mourey and McGinnis, 2003). In other words, the arrangement or distribution of facilities and activities in the surroundings of residential areas is one of the main factors found to influence urban transport patterns. Providing services and utilities for residents in proximity to their houses minimize the need to travel long distances and increase the chance of active travels. There has been a long tradition of investigating the association between the built environment and travel behaviour. Transport and urban planners have recently focused on promoting physical activity by environment-based solutions.

Pedestrian infrastructure including sidewalk access, quality and street connectivity have also been found as important criteria for determining walkability in neighbourhood areas, principally in micro-level studies (Lo, 2009). In some studies, these features have been found to affect comfort and safety of pedestrians (2004, Cervero and Duncan, 2003, Lo, 2009).

"Walk-Score" is one of the common approaches for measuring walkability. First introduced in 2007, it has been used in macro-level studies or when investigating land use features that affect proximity. The Walk Score algorithm considers points based on the distance to the closest facility in each land use category. In the closest facility in a category the distance ranges from 0.4 km to 1.6 km (REDFIN, 2015). In this approach facilities are categorised into offices, parks, theatres, schools and other common destinations. Duncan et al. (2011) and Carr et al. (2010) used walk-score in their study and claimed that walkability in neighbourhoods is based on the distance to different categories of services, including schools, parks and libraries.

Another common approach named Walkability Index (WI) introduced by Frank et al. (2007b, Frank et al., 2005, Frank et al., 2006, Frank et al., 2010). The walkability index is derived from four elements including dwelling density, street connectivity, land use mix and net retail areas. The walkability index is calculated from the sum of the z scores of the four mentioned measurements. The WI is one of the most common approaches used throughout the literature for measuring walkability (Frank et al., 2005, Frank et al., 2006, Frank et al., 2010, Peiravian et al., 2014, Giles-Corti et al., 2015, Sundquist et al., 2011, Owen et al., 2007a).

Although numerous studies have focused on measuring walkability, there has been limited research which has considered distance thresholds of walking to different destinations as one of the main barriers to active transport. Moreover the existing approaches may not answer the question of how far people are likely to walk to get to their desired destinations. Hence, this study describes a new concept to measure walking accessibility, the Walking Accessibility Index (WAI), a macro-level measurement, followed by an implementation of the new index in metropolitan Melbourne, Australia. This chapter compares the results of the WAI with those of one of the most common walkability approaches. The methodology section describes the approach used to compute the WAI, and the analysis and results of the application of the WAI in the Melbourne region, together with the results of the application of common existing approaches in Melbourne. The results of the comparison are then discussed, while in the closing section, conclusions and future directions of this study are outlined.

3.2 METHODOLOGY

Similar to previous chapters, WAI is also calculated for Melbourne's 9,510 Statistical Areas level 1 (SA1s) (Pink, 2011). Spatial Datasets used for the analysis were similar to ones used for CAI calculation. Followings briefly mentions the datasets, thereafter, the approach present a description of computation process.

3.2.1 Datasets

To calculate the WAI, the following datasets were utilised:

Points of Interest (POIs)

As explained in previous chapters, database of POIs was obtained from PSMA Australia (2011a), including urban centres, significant buildings, landmarks, public spaces, community facilities and indigenous locations, and included 15,588 points. For calculating the WAI, POIs were categorised into six groups of destinations including primary and secondary schools, tertiary institutions, child care centres, medical centres and retail and recreation centres.

Road Network Data

A dataset containing road networks (Swanson and McCormack, 2012) published by the Department of Environment, Land, Water & Planning was accessed. It contains line features delineating the state- wide road network, including bridges, connectors, footbridges, foot tracks and roads.

VISTA Dataset

The VISTA dataset (Transport, 2009), as described before, was provided from the Victorian Integrated Survey of Travel and Activity (VISTA). Dataset contained 17,089 walking trips. Travel data has been just used for evaluating the proposed index and not used in calculation process.

3.2.2 Approach

The aim of the present study is to measure the levels of walking accessibility for Melbourne's SA1s. For this purpose, weighted centroids of SA1s were considered as origins and POIs categories were defined as destinations. Using OD-cost matrix analysis, the average distances from each SA1 weighted centroid to all the available destinations were calculated. Travel impedance is defined based on the two thresholds of desirable and maximum travel distances. Origins and destinations and the approach are described in more detail in the following sections.

Origins (SA1s' Weighted Centroid)

As explained in previous sections weighted centroids of SA1s were obtained from the database of mesh blocks from the 2011 Census for the Melbourne Region was accessed from the ABS (ABS, 2011a).

Destinations (POI Categories)

POIs are categorised into six major destination groups, including primary and secondary schools, tertiary institutions, child care centres, medical centres, community services and libraries and retail and recreation centres. OD-cost matrix analysis was applied to each set of destinations separately. Two thresholds of distances, including the desirable and maximum travel times/distances, were adopted from the Austroads network operation planning framework (Espada et al., 2015, Espada and Luk, 2011). The Median Desirable Travel Distance (MDTD) is the value that satisfies half of the road users, while the Maximum Desirable Travel Distance (XDTD) is the value at which a significant percentage of people would find it unfeasible to regularly travel and they may be forced to relocate their residence closer to the destination or find a less suitable destination but one that is closer. The values considered for MDTD and XDTD are consistent with research conducted by Millward et al. the U.S.A. (Millward et al., 2013), Rattan et al. in Canada (Rattan et al., 2012), and Rendall et al. in New Zealand (Rendall et al., 2011). Table 1 shows MDTD and XDTD for destination categories.

Table 1. MDTD and XDTD for destination categories

Destination Categories	Thresholds*	
	MDTD	XDTD
Primary and Secondary Schools	< 800	< 1,600
Tertiary Institutions	< 1,200	< 2,400
Child Care Centres	< 800	< 1,600
Medical Centres	< 800	< 1,600
Retail and Recreation Centres	< 800	< 1,600
Community Services and Libraries	< 1,200	< 2,400

* Walking times were converted to distances assuming an average walking speed of 4.8 kilometres/hour or 80 meters/minutes (London, 2010).

Walking distances using network analysis by ArcGIS 10.2 were calculated for each SA1's weighted centroid to all available POIs within the acceptable travel distances. It should be noted that MDTD and XDTD were defined as cut-off values for each category. Then, average walking distances were computed for each centroid/origin.

WAI Calculation

WAI was calculated for each SA1 using the following formula:

$$WAI_{SA1_i} = \sum_{j=1}^{m} N_i \times \left(\frac{D_j^M - D_{ij}^A}{D_j^D} \right) \quad (1)$$

where, WAI_{SA1i} is the Walking Access Index for SA1 *i*, N_i is the number of POIs available within the acceptable walking distance, D_j^M is the maximum walking distance to destination type *j*, D_j^D denotes the desirable walking distance to destination type *j*, and D_{ij}^A represents the average walking distance from a *SA1* weighted centroid *i* to destination type *j*. The index can be grouped into six categories of accessibility levels, where category 1 represents a very poor level and level 6 represents an excellent level of accessibility (Table 3). A value of 0 indicates no accessibility in terms of the availability of destinations within the acceptable distance (cut-off value). The index reflects both the diversity and intensity of uses, while considering the availability of a number of destinations as well as the number of activities. A higher value of the WAI indicates a higher level of accessibility. Figure 4 in chapter 7 provides an example of the calculation of the WAI and shows how the value of the WAI changes for different levels of diversity and intensity of land use. More details and illustrations of WAI calculation is provided in a study by Saghapour et.al. (2017).

Ranges and categories of WAI are presented in Table 2. Values are grouped from very low to excellent level of access plus a zero group. The classification method used for the WAI categories is Quantile. This method simplifies comparison and improve the general map-reading (Brewer and Pickle, 2002). Zero accessibility is provided for 43,082 residents or 1.46% of SA1s. This category represents situations where there are either no destination groups or activities (POIs) within the walkable distances.

Table 2. WAI ranges and categories

WAI Categories	Ranges	Number of SA1s		Population	
		No.	%	No.	%
Zero/NA	0	139	1.46	43,082	1.10
Very Low	< 8	1,391	14.63	597,195	15.15
Low	8 – 14	1,563	16.44	652,103	16.55
Moderate	14 – 20	1,765	18.56	728,116	18.47
Good	20 – 25	1,122	11.80	456,400	11.58
Very good	25 – 37	1,891	19.88	780,475	19.80
Excellent	> 37	1,639	17.23	683,840	17.35
Total	–	9,510	100.00	3,941,211	100.00

Distribution of WAI categories is presented on the Figure 1. The WAI is categorized into six bands, where the first category represents very low walking accessibility while the last category signifies an excellent level of walking accessibility. The first and last categories have been further sub-divided into sub-levels to increase clarity. High levels of accessibility from good to excellent are mostly concentrated in the inner parts of the Melbourne region.

Table 3 summarize some descriptive statistics of the index elements. According to the contents of the table, average number of POIs per SA1 is 2.8. Average distances are also presented for each group of destinations. The output index (WAI) has an average of 24.1 with a maximum value of 221.4.

3.2.3 Walkability Index (WI)

WI is one of the most common approaches used for calculating walkability (Giles-Corti et al., 2015, Peiravian et al., 2014, Sundquist et al., 2011, Frank et al., 2010, Owen et al., 2007a, Frank et al., 2006, Frank et al., 2005). The typical form of the WI expression is as follows:

$$WI = \left(Zscore_{LUMIX}\right) + \left(Zscore_{Residential\_Density}\right) + \left(\alpha Zscore_{Connectivity}\right) \tag{2}$$

For being able to compare the proposed index (WAI) with the WI, we built WI values for SA1s. Hence, WIs were calculated for each SA1 as the sum of the z-scores for the three components included in the index, i.e. residential density (ratio of residential units to the residential area), street connectivity (intersection density), and land-use mix.

Figure 1. WAI categories within the Melbourne region

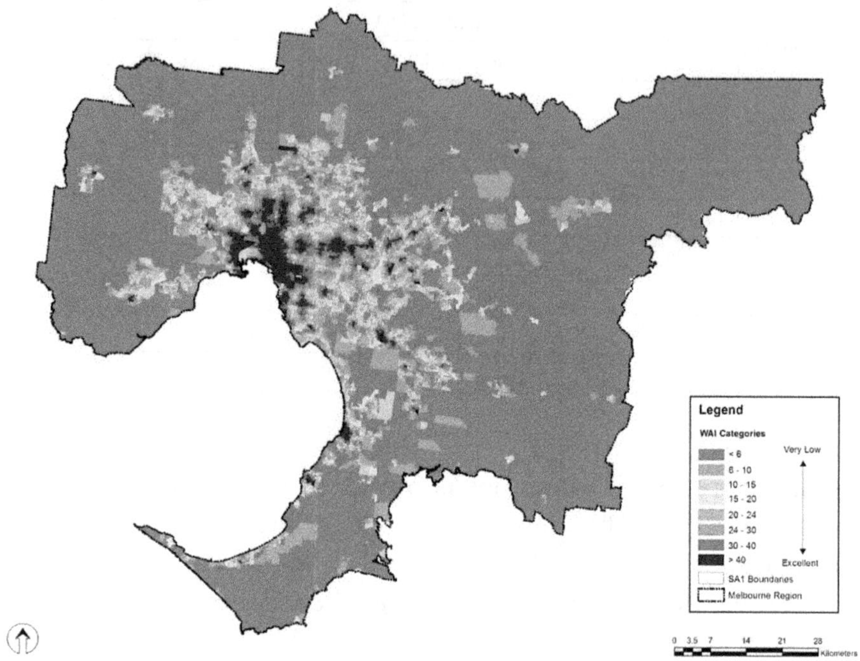

Table 3. Descriptive statistics of indicators of the index components

Indicators	Mean	Std. D	Min	Max
SA1's Area (km^2)	0.93	10.2	0.002	854.3
SA1's Population	414	209.5	0	6,224
Number of POIs per SA1	2.8	4.2	1	205
Distance of Primary and Secondary Schools	1,063.0	259.5	0	1,599.4
Distance of Tertiary Institutions	1,649.7	519.4	0	2,398.9
Distance of Child Care Centres	1,068.7	195.8	5.1	1,598.6
Distance of Medical Centres	1,111.7	272.4	17.2	1,599.6
Distance of Retail and Recreation centres	1,090.5	191.5	43.7	1,599.7
Distance of Community Services and Libraries	1,615.3	276.8	78.2	2,398.8
WAI	24.1	20.5	0	221.4

As shown in the equation 3. one of the components is the land-use mix, or entropy index (EI). This index indicates the degree to which a diversity of land-use types. For calculation of the EI, six different land use categories, including residential, commercial, industrial, transport and infrastructure, community services and sport and recreation centres, were chosen. These categories are adopted from ten main land use categories defined by the Australian Valuation Property Classification Codes (AVPCC) (Morse-McNabb, 2011).

$$EI = -\left(\sum_{j=1}^{J} \frac{P_j . \ln P_j}{\ln J}\right) \tag{3}$$

where, EI_i indicates the entropy index within a buffer *i*, P_j represents the proportion of land usc typc *j*, and *J* is thc number of land use categories. Values are normalised between 0 and 1, with 0 being single use and 1 indicating a completely even distribution of the six uses. The Australian Urban Research Infrastructure Network (AURIN) (Sinnott et al., 2011) developed the WI for neighbourhoods within the Melbourne region using Equation 3. The network provides a web-based environment for calculating WIs for different statistical subdivisions in the Melbourne area. This study applies the same method for calculating the WIs for SA1s. It should be noted that different studies consider different values for α as the coefficient for normalized values of connectivity. However, AURIN defines α as equal 1. The calculated WIs for SA1s varies from 1.8 to +50.8.

3.3 Data Analysis

For evaluating the proposed index, both WAI and WI were combined with VISTA dataset using the unique code of SA1. The VISTA dataset contains trip record information for 22,184 individuals from households randomly selected from 1,822 SA1s. The total number of trip stages reported by participants was 93,902, of which 17,089 were walking trips. The reason for using the trip stages for analysis is that walking trips are considered as the shortest, while covering all trip purposes including changing transport modes. Table 4 presents some descriptive statistics for the walking trips.

Table 4. Descriptive statistical measures of walking trips

Location		Variability	
Mean	26.796	Std Deviation	20.088
Median	22.000	Variance	403.516
Mode	8.000	Range	107.000
		Interquartile Range	26.000

*N= 16474 (Outliers were removed from analysis).

3.3.1 Measures of Association

Before applying the models, the strength of association between each of the indices and walking trip categories was examined. For this purpose, statistical measures of association were used for WAI/WI categories and walking trips. The results are presented in Table 5. The table indicates that the WAI has a stronger association with higher values of symmetric measures. The Somers' D, Gamma and Spearman tests are asymmetric measures of association between two variables, which plays a central role as a parameter in rank or non-parametric statistical methods (Newson, 2006). All the three tests ranged from -1.00 to 1.00, where 0 reflects no association, 1 reflects a positive and -1 indicates a negative perfect relationship between variables (Agresti and Kateri, 2011, Sprinthall, 2011). The following sections present the results of the models applied to the data while comparing the WAI with previous measurements.

3.3.2 Modelling and Interpretation

Models for this study were estimated using negative binomial regression (NBR) models (Saghapour, Moridpour, & Thompson, 2017). These kinds

Table 5. Tests of association between WAI/WI and walking trips

Symmetric Measures	WAI		WI	
	Stat.	p-Value	Stat.	p-Value
Somers' D	0.295	0.000	0.222	0.000
Gamma	0.365	0.000	0.275	0.000
Spearman Correlation	0.366	0.000*	0.286	0.000*
N of Valid Cases	16474	-	16474	-

* Significant at 0.99 confidence level

of models, as explained in the first chapter, are usually run for count data and require positive integers for the dependent variable. Since the number of trips is always a non-negative integer, this study adopted the NBR regression technique (Coruh et al., 2015). For more information please refer to chapter 1 section 1.4.2.

In NBR models, it is assumed that variance is bigger than the mean (Cao et al., 2006). In other words, NBR models capture the over-dispersion (variance > mean). Hence, to check this assumption the histogram of walking trips has been examined. As shown in the figure walking trips used in this study follow this pattern (see Figure 2).

Two separate NBR models were generated for walking trips using a different walkability measurement (WAI and WI) in every run, while keeping other variables constant in the model. M1 presents the results of a NBR model considering all the predictor variables and the WAI, and M2 denotes the model with all the variables used in M1 while replacing WI for the walkability measure. Walking trips are defined as a count-dependent variable. Age, gender, car licence, employment type, household size (HHS), household structure (HHSR), and the number of cars in the households were employed as socioeconomic variables (Lee et al., 2014, Jun et al., 2012, Shay

Figure 2. Histogram of walking trips

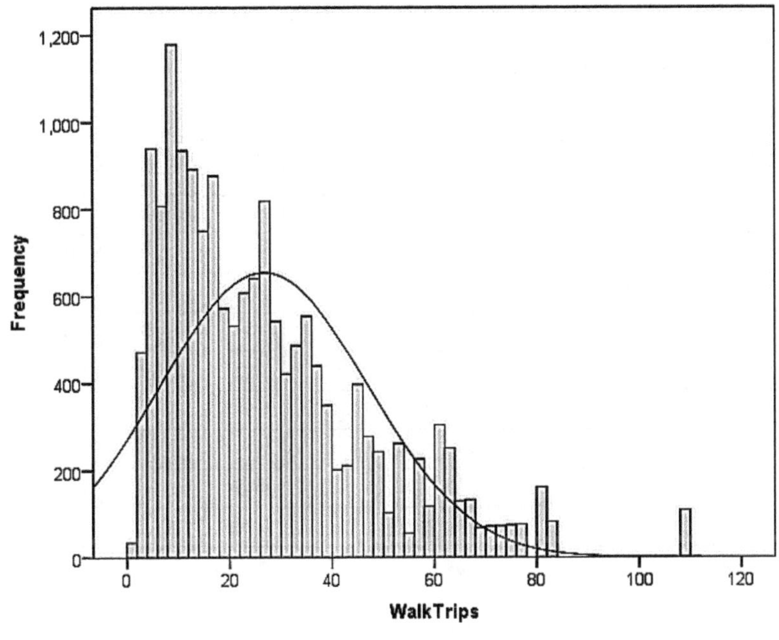

Table 6. Independent variables and their expected associations with walking trips

Variables	Description
Socio-demographic Age	Age of respondent
Sex	Gender
LNC	Driver licence
WT	Type of work
HHS	Usual number of residents in household
HHSR	Demographic structure of household
Car	Number of cars in the household
Built Environment WAI	Walking Access Index
WI	Walkability Index

Note: HHSR is converted to five dummy variables: sole person, couple no children, couple with children, one parent and other. WT is converted into five dummy variables: full- time, part- time, casual, unemployed and not working; sex and LNC are defined as binary variables.

and Khattak, 2012, Ewing and Cervero, 2010, Winters et al., 2010). Table 6 shows the list of independent variables and their description.

Table 7 presents the descriptive statistics for the variables used in the models. These statistics were calculated for 16,474 records of trip stages excluding the outliers. According to the content of the table, the average age of the respondents was 38 years old and respondent were equally distributed

Table 7. Descriptive statistics of variables

Variable	Mean	S.D.	Min	Max
Walking Trips	26.80	20.09	1.00	108.00
Age	36.88	19.31	0.00	90.00
Sex	1.54	0.50	1.00	2.00
LNC	1.29	0.45	1.00	2.00
HHS	3.00	1.37	1.00	6.00
WT	2.89	1.78	1.00	5.00
HHSR	2.80	1.13	1.00	5.00
Car	1.78	0.85	1.00	4.00
WAI	31.35	18.94	0.00	109.63
WI	0.61	1.76	-1.78	12.42

n=16,474 trip stages

in terms of gender. The average usual number of residents in households was about three residents.

In order to examine the practicality of the WAI compared to the WI, two NBR models were run. WAI was included in M1 along with all other variables and WI was replaced with WAI having all other variable same as M1 (see Table 8). The NBR models predicted walking trips with age, gender, licence, work type, HHSR, HHS, car ownership, and walkability measurements. Except for gender, couple with children and couple without children, other variables were statistically significant. As shown in Table 9, the dispersion parameters of the models are greater than zero (about 0.5), which indicates that the response variable is over-dispersed, hence the NBR model was found to be more appropriate for the data. If the dispersion parameter equals zero, the model reduces to the simpler Poisson model (Hilbe, 2011).

The Incident Rate Ratio (IRR) in NBR models describes the percentage change in the response variable for every unit increase in the corresponding explanatory variable (Hilbe, 2008). Therefore, according to the results, there is a 33% increase in walking trips for every unit increase in WAI, while this number is 25% for WI. As the age increases by one unit, walking trips decreases by 1%. There is a 10% decrease in walking trips by one unit increase in number of cars in the household. People with part-time jobs have 12% more walking trips than those who are not working. People who live alone have 13% fewer walking trips than others.

According to Table 8, the Value/DFs for M1 and M2 were 1.0679 and 1.0706, respectively. That shows both models were fitted on data well. Furthermore, M1 has the lowest Akaike Information Criterion (AIC) and Bayesian Information Criterion (BIC), which are measures of the relative quality of statistical models for a given set of data. Given a series of models for the data, these criteria estimate the quality of each model, relative to each of the other models (Boisbunon et al., 2014, Hu, 2007, Aho et al., 2014).

In this study, M1 had the smaller values for both AIC and BIC that indicate M1 was better fitted on data compared to M2. The estimated coefficients from the models were compared with each other, and the results are presented in Table 8. The t-statistics results indicate that there is a significant difference between the coefficients of walking accessibility measurements estimated by the two models.

Table 8. Outputs of the NB regression model for walking trips

Parameter	M1				M2			
	Estimate	IRR	Std. Error	Wald Chi-Square	Estimate	IRR	Std. Error	Wald Chi-Square
Intercept	3.2946***	-	0.0388	7228.51	3.2408***	-	0.0393	6804.02
Age	-0.0008**	0.9992	0.0004	4.36	-0.0008**	0.9992	0.0004	3.94
Sex (Male)	-0.0134	0.9867	0.012	1.25	-0.0224*	0.9778	0.0122	3.4
LNC (Yes)	0.0773***	1.0804	0.0189	16.74	0.0738***	1.0766	0.0192	14.85
HHS	0.0300***	1.0304	0.0074	16.31	0.0433***	1.442	0.0075	33.05
WT								
Full Time	0.0504**	1.0517	0.0163	9.62	0.0731***	1.0759	0.0165	19.67
Part Time	0.1162***	1.1233	0.0198	34.58	0.117***	1.1242	0.0201	33.86
Casual	0.0486**	1.0498	0.0249	3.83	0.0802**	1.0835	0.0252	10.14
Unemployed	0.106**	1.1118	0.0474	5.01	0.2055***	1.2281	0.048	18.33
HHSR								
Sole Person	-0.1346***	0.8740	0.0326	17.08	-0.0309	0.9695	0.0329	0.89
Couple with kids	0.0048	1.0045	0.0247	0.04	0.0299	1.0303	0.0249	1.44
Couple without kids	-0.0019	0.9981	0.0211	0.01	0.0102	1.0103	0.0215	0.22
Single parent	-0.2031***	0.8162	0.0288	49.66	-0.1992***	0.8194	0.0292	46.5
Car	-0.0966***	0.9079	0.008	145.08	-0.1099***	0.8959	0.008	186.65
WAI	0.2879***	1.3337	0.0073	1567.77	-	-	-	-
WI	-	-	-	-	0.2240***	1.2511	0.0068	1098.17
Dispersion	0.4706	-	0.0058	-	0.4856	-	0.0059	-

Notes: (1) number of walking trips is defined as a dependant variable.

(2) To be able to compare the walkability indexes with different measurement scales, WAI and WI were standardized. The dependent variable was not standardized, since NBR requires the dependent variable to be a count value (non-negative integer).

(3) The NB dispersion parameter was estimated by maximum likelihood.

(4) For WAI and WI z-values of the variables used in the model.

(5) Signifiance codes: $p < 0.001$ '***', 0.01 '**', 0.1 '*'.

(6) Overall goodness-of-fit:

M1: Value/DF = 1.0679; AIC = 119393.18; BIC = 119514.44,

M2: Value/DF = 1.0706; AIC = 119860.78; BIC = 119982.04.

3.4 DISCUSSIONS AND CONCLUSION

This chapter presented the results of a new approach proposed for measuring walkability within metropolitan areas. Presented approach can be employed by planners and policy makers to compare and rank areas already built, and identify new areas where investment might improve walking accessibility.

The method used for calculating WAI not only considers the proximity of different uses, but also considered the number of activities within different destination categories. As Dong et al. (Dong et al., 2016) argued walking can be enhanced by improving the quality of the built environment. This can affect travel walking time/distance and transport mode choice. In urban and transport planning, much effort is currently being put into providing friendly environments to encourage walking in cities. According to Peiravian et al. (2014), measuring the friendliness of neighbourhoods as a policy tool to promote more walking and cycling remains important, and requires more research. This study provided a starting point for such a task.

In summary, this paper introduced a new approach measuring walking accessibility. The techniques presented are easy and simple to apply while they can be used for any geographical scales. WAI measured both diversity and intensity of the land uses within the defined areas.

The results indicated that 1.5% of SA1s representing 1.1% of Melbourne population have no walking access to different destinations, while 15.2% and 16.6% of residents have very low or low walking access. The inner area of Melbourne covers approximately 3,504 km2, of which approximately 1,457 km2 (42% of the inner area) is covered by zero to moderate levels of walking accessibility. These numbers imply that a considerable number of SA1s have a low to average levels of walking accessibility. These findings signify the high concentration of POIs in the inner part of Melbourne and the CBD, and that the inner areas of Melbourne have better walking access than the outer areas. In addition, the outer suburbs are characterised by dispersed patterns, which may result in increasing the distances and decreasing the odds of walking.

For assessing and evaluating the WAI, one of the most common approaches for measuring walkability, WI, was generated for 9510 statistical areas (SA1s). Thereafter, both indexes were joined to the VISTA dataset using the SA1s' unique code. Tests of association were generated to examine whether there is a stronger relationship between the new index and the number of walking trips compared to the existing WI. These findings show that the association values for WAI both in ordinal and interval tests were higher than those for the WI. For assessing the practicality of the proposed index, WAI and WI along with a series of socioeconomic characteristics, were employed in two separate NBR models. M1 model included the WAI with other predictor variables, whilst the M2 model used the WI as the measure of walking accessibility. Comparison of the results revealed that M1 had the lowest AIC (AICM1 = 119393.18 < AICM2 = 119860.78) and BIC (BICM1 = 119514.44 < BICM2 = 119982.04), which showed a better fit for the data. The IRR for

WAI in M1 (IRRWAI=1.33) was higher than the coefficients estimated for WI (IRRWI=1.25) in M2. These figures indicated that more walking trips are expected when there is a one-unit increase in the WAI compared to the WI. Therefore, WAI is evaluated as a valid means of measuring walkability in the Melbourne region based on the VISTA database.

3.5 FUTURE RESEARCH DIRECTIONS

The literature commonly reports that built environment features such as density, diversity, and road connectivity can promote walking trips. Current study mainly focused on investigating whether distance thresholds overcome features considered in other measures, such as connectivity and/or urban design factors. The results of the analysis revealed that people are more likely to walk when their desired destination is located within the distance thresholds. In terms of numbers of walking trips, the findings show that the average number of walking trips within SA1s (the second smallest of Melbourne's geographical areas) is higher when WAI is higher.

The techniques presented are straightforward to apply. The WAI shows greater accuracy than the WI for measuring walkability based on the VISTA dataset. The quantitative approach is designed to be applied with available census data and network modelling tools. Furthermore, the analysis provides reliable and defendable results, which can be computed for 98.5% of SA1s.

One of the limitations of the study is that several of the categories are likely to be single buildings, such as child care centres and libraries, while other categories such as retail centres, are collections of multiple shops which may show stronger attraction. An additional limitation is that, as many authorities are not likely to have a similar POI database, widespread use may be limited. However, as long as land-use maps are available, a POI database can be created by turning features into points. Hence, future work may consider these points.

REFERENCES

ABS. (2011). *Australian Bureau of Statistics*. Canberra, Australia: Australia, Year Book.

Agresti, A., & Kateri, M. (2011). *Categorical data analysis*. Springer.

Aho, K., Derryberry, D., & Peterson, T. (2014). Model selection for ecologists: The worldviews of AIC and BIC. *Ecology*, *95*(3), 631–636. doi:10.1890/13-1452.1 PMID:24804445

Bauman, A. E., Reis, R. S., Sallis, J. F., Wells, J. C., Loos, R. J., Martin, B. W., & Group, L. P. A. S. W. (2012). Correlates of physical activity: Why are some people physically active and others not? *Lancet*, *380*(9838), 258–271. doi:10.1016/S0140-6736(12)60735-1 PMID:22818938

Boisbunon, A., Canu, S., Fourdrinier, D., Strawderman, W., & Wells, M. T. (2014). Akaike's information criterion, Cp and estimators of loss for elliptically symmetric distributions. *International Statistical Review*, *82*(3), 422–439. doi:10.1111/insr.12052

Brewer, C. A., & Pickle, L. (2002). Evaluation of methods for classifying epidemiological data on choropleth maps in series. *Annals of the Association of American Geographers*, *92*(4), 662–681. doi:10.1111/1467-8306.00310

Cao, X., Handy, S. L., & Mokhtarian, P. L. (2006). The influences of the built environment and residential self-selection on pedestrian behavior: Evidence from Austin, TX. *Transportation*, *33*(1), 1–20. doi:10.100711116-005-7027-2

Carr, L. J., Dunsiger, S. I., & Marcus, B. H. (2010). Walk score™ as a global estimate of neighborhood walkability. *American Journal of Preventive Medicine*, *39*(5), 460–463. doi:10.1016/j.amepre.2010.07.007 PMID:20965384

Cervero, R., & Duncan, M. (2003). Walking, bicycling, and urban landscapes: Evidence from the San Francisco Bay Area. *American Journal of Public Health*, *93*(9), 1478–1483. doi:10.2105/AJPH.93.9.1478 PMID:12948966

Coruh, E., Bilgic, A., & Tortum, A. (2015). Accident analysis with aggregated data: The random parameters negative binomial panel count data model. *Analytic Methods in Accident Research*, *7*, 37–49. doi:10.1016/j.amar.2015.07.001

Dannenberg, A. L., Jackson, R. J., Frumkin, H., Schieber, R. A., Pratt, M., Kochtitzky, C., & Tilson, H. H. (2003). The impact of community design and land-use choices on public health: A scientific research agenda. *American Journal of Public Health*, *93*(9), 1500–1508. doi:10.2105/AJPH.93.9.1500 PMID:12948970

Dong, H., Ma, L., & Broach, J. (2016). Promoting sustainable travel modes for commute tours: A comparison of the effects of home and work locations and employer-provided incentives. *International Journal of Sustainable Transportation, 10*(6), 485–494. doi:10.1080/15568318.2014.1002027

Duncan, D. T., Aldstadt, J., Whalen, J., Melly, S. J., & Gortmaker, S. L. (2011). Validation of Walk Score® for estimating neighborhood walkability: An analysis of four US metropolitan areas. *International Journal of Environmental Research and Public Health, 8*(11), 4160–4179. doi:10.3390/ijerph8114160 PMID:22163200

Espada, I., Bennett, P., Green, D., & Hatch, D. (2015). *Development of the accessibility-based network operations planning framework*. Academic Press.

Espada, I. & Luk, J. (2011). *Application of accessibility measures*. Academic Press.

Ewing, R., & Cervero, R. (2010). Travel and the built environment. *Journal of the American Planning Association, 76*(3), 265–294. doi:10.1080/01944361003766766

Ewing, R., Schmid, T., Killingsworth, R., Zlot, A., & Raudenbush, S. (2003). Relationship between urban sprawl and physical activity, obesity, and morbidity. *American Journal of Health Promotion, 18*(1), 47–57. doi:10.4278/0890-1171-18.1.47 PMID:13677962

Frank, L., Engelke, P., & Schmid, T. (2003). *Health and community design: The impact of the built environment on physical activity*. Island Press.

Frank, L. D., Andresen, M. A., & Schmid, T. L. (2004). Obesity relationships with community design, physical activity, and time spent in cars. *American Journal of Preventive Medicine, 27*(2), 87–96. doi:10.1016/j.amepre.2004.04.011 PMID:15261894

Frank, L. D., Sallis, J. F., Conway, T. L., Chapman, J. E., Saelens, B. E., & Bachman, W. (2006). Many pathways from land use to health: Associations between neighborhood walkability and active transportation, body mass index, and air quality. *Journal of the American Planning Association, 72*(1), 75–87. doi:10.1080/01944360608976725

Frank, L. D., Sallis, J. F., Saelens, B. E., Leary, L., Cain, K., Conway, T. L., & Hess, P. M. (2010). The development of a walkability index: Application to the Neighborhood Quality of Life Study. *British Journal of Sports Medicine*, *44*(13), 924–933. doi:10.1136/bjsm.2009.058701 PMID:19406732

Frank, L. D., Schmid, T. L., Sallis, J. F., Chapman, J., & Saelens, B. E. (2005). Linking objectively measured physical activity with objectively measured urban form: Findings from SMARTRAQ. *American Journal of Preventive Medicine*, *28*(2), 117–125. doi:10.1016/j.amepre.2004.11.001 PMID:15694519

Giles-Corti, B., Macaulay, G., Middleton, N., Boruff, B., Bull, F., Butterworth, I., ... Christian, H. (2015). Developing a research and practice tool to measure walkability: A demonstration project. *Health Promotion Journal of Australia*, *25*(3), 160–166. doi:10.1071/HE14050 PMID:25481614

Hayashi, T., Tsumura, K., Suematsu, C., Okada, K., Fujii, S., & Endo, G. (1999). Walking to work and the risk for hypertension in men: The Osaka Health Survey. *Annals of Internal Medicine*, *131*(1), 21–26. doi:10.7326/0003-4819-131-1-199907060-00005 PMID:10391811

Hilbe, J. M. (2008). *Brief overview on interpreting count model risk ratios: An addendum to negative binomial regression*. Cambridge University Press.

Hilbe, J. M. (2011). *Negative binomial regression*. Cambridge University Press. doi:10.1017/CBO9780511973420

Hoogendoorn, S. P., Bovy, P., & Daamen, W. (2004). Walking infrastructure design assessment by continuous space dynamic assignment modeling. *Journal of Advanced Transportation*, *38*(1), 69–92. doi:10.1002/atr.5670380106

Hu, S. (2007). *Akaike information criterion*. Center for Research in Scientific Computation.

Jun, M.-J., Kim, J. I., Kwon, J. H., & Jeong, J.-E. (2012). The effects of high-density suburban development on commuter mode choices in Seoul, Korea. *Cities (London, England)*.

Lavizzo-Mourey, R., & Mcginnis, J. M. (2003). Making the case for active living communities. *American Journal of Public Health*, *93*(9), 1386–1388. doi:10.2105/AJPH.93.9.1386 PMID:12948948

Lee, J.-S., Nam, J., & Lee, S.-S. (2014). Built environment impacts on individual mode choice: An empirical study of the Houston-Galveston metropolitan area. *International Journal of Sustainable Transportation*, *8*(6), 447–470. doi:10.1080/15568318.2012.716142

Lo, R. H. (2009). Walkability: What is it? *Journal of Urbanism*, *2*(2), 145–166. doi:10.1080/17549170903092867

London, T. F. (2010). *Measuring Public Transport Accessibility Levels.* Academic Press.

Manaugh, K., & El-Geneidy, A. (2011). Validating walkability indices: How do different households respond to the walkability of their neighborhood? *Transportation Research Part D, Transport and Environment*, *16*(4), 309–315. doi:10.1016/j.trd.2011.01.009

Manson, J. E., Hu, F. B., Rich-Edwards, J. W., Colditz, G. A., Stampfer, M. J., Willett, W. C., ... Hennekens, C. H. (1999). A prospective study of walking as compared with vigorous exercise in the prevention of coronary heart disease in women. *The New England Journal of Medicine*, *341*(9), 650–658. doi:10.1056/NEJM199908263410904 PMID:10460816

Millward, H., Spinney, J., & Scott, D. (2013). Active-transport walking behavior: Destinations, durations, distances. *Journal of Transport Geography*, *28*, 101–110. doi:10.1016/j.jtrangeo.2012.11.012

Morse-Mcnabb, E. (2011). The Victorian Land Use Information System (VLUIS): A new method for creating land use data for Victoria, *Australia. Surveying and Spatial Sciences Conference*, 155.

Newson, R. (2006). Confidence intervals for rank statistics: Somers' D and extensions. *The Stata Journal*, *6*(3), 309–334. doi:10.1177/1536867X0600600302

Owen, N., Cerin, E., Leslie, E., Coffee, N., Frank, L. D., Bauman, A. E., ... Sallis, J. F. (2007a). Neighborhood walkability and the walking behavior of Australian adults. *American Journal of Preventive Medicine*, *33*(5), 387–395. doi:10.1016/j.amepre.2007.07.025 PMID:17950404

Owen, N., Cerin, E., Leslie, E., Dutoit, L., Coffee, N., Frank, L. D., ... Sallis, J. F. (2007b). Neighborhood walkability and the walking behavior of Australian adults. *American Journal of Preventive Medicine*, *33*(5), 387–395. doi:10.1016/j.amepre.2007.07.025 PMID:17950404

Peiravian, F., Derrible, S., & Ijaz, F. (2014). Development and application of the Pedestrian Environment Index (PEI). *Journal of Transport Geography, 39*, 73–84. doi:10.1016/j.jtrangeo.2014.06.020

Pink, B. (2011). Australian statistical geography standard (ASGS): volume 5–remoteness structure. Canberra: Australian Bureau of Statistics.

PSMA. (2011). *2012 Annual report*. Canberra, Australia: Mapping Data for Australia.

Rattan, A., Campese, A. & Eden, C. (2012). Modeling walkability. *Arc. User. Winter, 2012*, 30-3.

REDFIN. (2015). *How Walk Scor Works*. Available: https://www.redfin.com

Rendall, S., Page, S., Reitsma, F., Van Houten, E., & Krumdieck, S. (2011). Quantifying transport energy resilience: Active mode accessibility. *Transportation Research Record: Journal of the Transportation Research Board, 2242*(1), 72–80. doi:10.3141/2242-09

Saelens, B. E., Sallis, J. F., Black, J. B., & Chen, D. (2003). Neighborhood-based differences in physical activity: An environment scale evaluation. *American Journal of Public Health, 93*(9), 1552–1558. doi:10.2105/AJPH.93.9.1552 PMID:12948979

Saghapour, T., Moridpour, S., & Thompson, R. G. (2017). Estimating Walking Access Levels Incorporating Distance Thresholds of Built Environment Features. *International Journal of Sustainable Transportation*, 1–14. doi:10.1080/15568318.2017.1380245

Saghapour, T., Moridpour, S., & Thompson, R. G. (2017). Measuring Walking Accessibility in Metropolitan Areas. *Transportation Research Record, Journal of the Transportation Research Board, 2661*, 111-119.

Shay, E., & Khattak, A. J. (2012). Household travel decision chains: Residential environment, automobile ownership, trips and mode choice. *International Journal of Sustainable Transportation, 6*(2), 88–110. doi:10.1080/15568318.2011.560363

Sinnott, R., Galang, G., Tomko, M. & Stimson, R. (2011). *Australian Urban Research Infrastructure Network*. Academic Press.

Sprinthall, R. C. (2011). *Basic statistical analysis*. Academic Press.

Sundquist, K., Eriksson, U., Kawakami, N., Skog, L., Ohlsson, H., & Arvidsson, D. (2011). Neighborhood walkability, physical activity, and walking behavior: The Swedish Neighborhood and Physical Activity (SNAP) study. *Social Science & Medicine*, *72*(8), 1266–1273. doi:10.1016/j.socscimed.2011.03.004 PMID:21470735

Swanson, K. C., & Mccormack, G. R. (2012). The Relations Between Driving Behavior, Physical Activity, and Weight Status Among Canadian Adults. *Journal of Physical Activity & Health*, *9*(3), 352–359. doi:10.1123/jpah.9.3.352 PMID:21934155

Transport, D. O. (2009). *Victorian Integrated Survey of Travel and Activity*. Available: http://www5.transport.vic.gov.au/

Winters, M., Brauer, M., Setton, E. M., & Teschke, K. (2010). Built environment influences on healthy transportation choices: Bicycling versus driving. *Journal of Urban Health*, *87*(6), 969–993. doi:10.100711524-010-9509-6 PMID:21174189

Chapter 4
Applying Decision Tree Approaches on Vehicle-Pedestrian Crashes

ABSTRACT

In the Melbourne metropolitan area in Australia, an average of 34 pedestrians were killed in traffic accidents every year between 2004 and 2013, and vehicle-pedestrian crashes accounted for 24% of all fatal crashes. Mid-block crashes accounted for 46% of the total pedestrian crashes in the Melbourne metropolitan area and 49% of the pedestrian fatalities occurred at mid-blocks. Many studies have examined factors contributing to the frequency and severity of vehicle-pedestrian crashes. While many of the studies have chosen to focus on crashes at intersections, few studies have focussed on vehicle-pedestrian crashes at mid-blocks. Since the factors contributing to vehicle crashes at intersections and mid-blocks are significantly different, more research needs to be done to develop a model for vehicle-pedestrian crashes at mid-blocks. In order to identify factors contributing to the severity of vehicle-pedestrian crashes, three models using different decision trees (DTs) were developed. To improve the accuracy, stability, and robustness of the DTs, bagging and boosting techniques were used in this chapter. The results of this study show that the boosting technique improves the accuracy of individual DT models by 46%. Moreover, the results of boosting DTs (BDTs) show that neighbourhood social characteristics are as important as traffic and infrastructure variables in influencing the severity of pedestrian crashes.

DOI: 10.4018/978-1-5225-7943-4.ch004

4.1. INTRODUCTION

Walking is the most basic and active mode of travel in transportation systems. In order to reduce air pollution and obtain better public health outcomes, efforts to encourage non-motorized transport modes have increased in recent years (Wey & Chiu 2013). To increase the number of walking trips, concerns about pedestrian safety must be addressed. Pedestrians are more likely to be harmed or killed in traffic crashes. They are 23 times more likely to be killed than vehicle occupants (Miranda-Moreno et al, 2011) and more than 22% of traffic deaths in the world are of pedestrians (WHO 2013). Every year, 34 pedestrians are killed in traffic crashes in the Melbourne metropolitan area, representing 24% of the total traffic fatalities. Mid-block crashes account for 46% of total pedestrian crashes in the Melbourne metropolitan area and 49% of pedestrian fatalities occur at mid-blocks (Crash Statistics Data 2016).

Many studies have been conducted to examine the factors contributing to the frequency and severity of vehicle-pedestrian crashes (Anderson et al, 1997; Zajac & Ivan, 2003; Kim et al, 2008; Tulu et al, 2015). Whereas many of the studies have chosen to focus on crashes at intersections (Lee & Abdel-Aty, 2005), only a few studies have chosen to focus on vehicle-pedestrian crashes at mid-blocks. Since the factors contributing to vehicle crashes at intersections and mid-blocks are significantly different (Al-Ghamdi, 2003; Bennt & Yiannakoulias, 2015), more research needs to be done to develop a model for vehicle-pedestrian crashes at mid-blocks. In terms of the methodologies used to analyse vehicle-pedestrian crashes, our review of the literature shows that different regression techniques, such as logit and probit models, are widely used. However, these statistical models require specific assumptions on the distribution of the random term and the relationship between the dependent and independent variables (Chang & Wang 2006). To circumvent these restrictions, decision trees (DTs) have been increasingly used in road safety studies (Lord et al, 2007). One disadvantage of this approach is that the results obtained in standard DTs may be changed significantly with changes in training and testing the data (Lord et al, 2007). To increase stability and robustness, ensemble methods, such as bagging and boosting, have recently been used in some traffic safety studies (Abdelwahab & Abdel-Aty, 2001, Zajac & Ivan 2003, Lefler & Gabler 2004, Chong et al. 2005). However, the relative performance of these methods has yet to be investigated.

The main objective of this research is to identify the factors contributing to the severity of vehicle-pedestrian crashes at mid-blocks. Whereas previous studies have mainly focussed on pedestrian crashes at intersections or examined the crash risk at mid-blocks for special groups of pedestrians (e.g. children) or specific study areas (e.g. pedestrian crossings), this research will examine all vehicle-pedestrian crashes in mid-blocks in the Melbourne metropolitan area. In addition, this research will consider different socio-economic variables, such as place of birth, level of education and percentage of labour force participation in a neighbourhood or surrounding suburbs. The distance of the crash location to/from public transport stops is another variable used in this research as a novel measure to identify the influence of public transport stops on pedestrian crashes. Furthermore, this study will compare the performance of a single DT model with a boosted and a bagged DT models. These three models have been applied in different studies to explore the factors contributing to vehicle crashes. However, this is the first time that these models have been developed for pedestrian crashes. Moreover, partial dependence plotting is used in this research for the first time in traffic crash studies. Partial dependence plots depict the relationship between the severity levels and one predictor variable while considering the average effects of all other predictors.

This chapter is structured as follows. The next section of the chapter provides a review of the previous studies on pedestrian crash severity modelling. Section 4.3 describes the data used in the study, while Section 4.4 presents the methodology of this research. The results are presented and discussed in Section 4.5. Finally, the outcomes are summarised and directions for future research are presented in Section 4.6.

4.2. LITERATURE REVIEW

Improving pedestrian safety requires comprehensive exploration and analysis of the factors influencing the probability of pedestrian crash occurrence and pedestrian crash severity levels. According to the literature, many studies try to find the impact of specific factors (such as pedestrian age, speed, light condition, etc.) on pedestrian crashes. In addition, there are many studies that develop a specific model (such as binary models, ordered discrete models and unordered multinomial discrete models) to find risk factors affecting the severity of pedestrian crashes. For instance, many studies evaluated the impact of pedestrian age on crash severity level demonstrating that pedestrian age can significantly impact crash severity (Lee and Abdel-Aty, 2005, Eluru et al, 2008, Kim et al, 2008, Kim et al, 2010a, Kim et al, 2010b, Sarkar et al, 2011, Oikawa et al, 2016).

These studies indicated that pedestrian crash severity rises by increase in pedestrian age. In addition, a number of studies attempted to identify the impacts of driver gender, age and alcohol consumption on crash injury severity levels (Miles-Doan, 1996, Laflamme et al. 2005, Lee and Abdel-Aty, 2005, Kim et al, 2008, Kim et al, 2010). Female pedestrians have found to be the most vulnerable group in pedestrian crashes. For instance, Lee and Abdel-Aty (2005) found that female pedestrians have higher severity levels than male pedestrians. Also, Miles-Doan (1996) and Kim et al. (2008, 2010) showed that drinking and driving, and pedestrian alcohol consumption can significantly increase the risk of pedestrian fatal crashes.

Also, some studies have focused on the effects of traffic control type on crash severity levels (Lee and Abdel-Aty, 2005, Eluru et al, 2008). The findings indicated that pedestrian crash injury levels increase in absence of traffic control such as traffic signals, signs or pedestrian signals. Furthermore, several studies examined the impacts of vehicle type (Ballesteros et al, 2004, Newstead & D'Elia, 2010, Aziz et al, 2013), weather condition (Kim et al, 2010b), and road speed limit (Sasidharan & Mene'ndez, 2014) on pedestrian crash injury severity levels.

According to the literature, pedestrian studies mainly considered crashes at intersections. However, in recent years mid-block crashes have been examined in some research. For instance, Quistberg et al. (2015) applied multilevel models to estimate the risk of pedestrian crashes at intersections and mid-blacks in Seattle, the U.S. In another study, Zheng et al. (2015) modelled the interaction between pedestrian behaviours such as gap acceptance and speed at intersections and mid-blocks. In this study, they used the data that is collected from pedestrian crossing roads on the campus of University of Florida. In addition, Bennet and Yiannakoulias (2015) applied a conditional logistic regression to predict the log-odds of child pedestrian collision risk at intersections and mid-blocks in Hamilton, Ontario, Canada. They used limited road condition variables such as existence of bus and bike lane, speed limit, sidewalk and land use characteristics to predict the risk of accidents for child pedestrians.

Considering the model type, different statistical approaches have been applied to analyse pedestrian crash injury severities. Review of the literature showed that the binary models, ordered discrete models and unordered multinomial discrete models are three main statistical techniques that have been widely used to study pedestrian crash severity levels. In binary crash severity models, outcomes are injury against non-injury crashes or fatal against non-fatal crashes. These studies have commonly used discrete models such as binary Logit and binary Probit models. In recent studies, Koepsell et al. (2002) and Moudon et al. (2011) developed a logistic regression model for pedestrian crashes to analyse severity for this type of crashes.

According to the ordinal nature of crash severity levels, ordered probability models are very popular in traffic crash studies. Lee and Abdel-Aty used this approach to estimate the likelihood of pedestrian injury severity at intersections (Cui & Nambisan, 2003). In ordered logit and ordered probit models, it is assumed that the parameter estimates are constant across different severity levels. However, some covariates might increase the probability of one type of crash severity level in practice; while they might decrease the probability of occurrence of other severity levels (Savolainen et al, 2011). To deal with this limitation of ordered logit models, Eluru et al. (2008) developed a generalized ordered probability model to examine the crash injury severity levels of pedestrians and bicyclists in U.S. In this model, they allowed the thresholds in ordered probability model to vary based on both observed and unobserved characteristics.

Limitations of the traditional ordered Logit and Probit models lead to developing unordered models to analyse traffic crash injury severity levels. Multinomial Logit model (MNL), Mix Logit model and random parameter Logit model are the most common unordered models that are applied in many pedestrian crash studies (Wier et al, 2009, Siddiqui et al, 2012, Yingying et al, 2012). For instance, Kim et al. (2010a) used mixed logit models for pedestrian crashes to identify risk factors that increase the probability of fatal and serious injuries for this road user group. Light condition, road type, speed limit, and driver alcohol use all play important roles in determining the crash severity levels. However, these models have similar limitations to logit models. For instance, all explanatory variables must be independent to each other.

Machine learning is another approach that is widely used in different areas of civil engineering (Reich, 1997, Hung & Jan, 1999), such as construction and structure design, pavement design and traffic engineering (Adeli & Balasubramanyam 1988, Thurston & Sun, 1994, Herabat & Songchitruksa, 2003, Aghabayk et al, 2013, Celikoglu 2013). In recent years, non-parametric techniques have become popular and have been used in traffic crash severity modelling. Kuhnert et al. (2000) applied Classification And Regression Trees (CART) and Multivariate Adaptive Regression Splines (MARS) to estimate motor vehicle injuries. They showed that the combined use of MARS and CART can be a useful method to display more detailed analysis compared to traditional methods such as logistic regression. In another work, Chong et al. (2015) and Yu et al. (2014) used different machine learning paradigms to model the traffic crash severity. In that study, Neural Networks was trained using hybrid learning approaches, support vector machines and DTs. Then, concurrent hybrid models using DTs and neural networks have been developed. In other works, CART model was applied to analyse the traffic crash data and find influencing variables on traffic injury and fatal accidents (Kashani & Mohaymany, 2011, Abellán et al, 2013, Chung, 2013, Chang & Chien, 2013, De Oña J et al, 2013, Peña-garcía et al, 2014, Kwon et al, 2015, Wang et al, 2015). For example, Chang and Chien (2013) applied CART model to explore the relationship between accident injury severity levels and driver/vehicle characteristics, highway geometric variables, environmental characteristics, and accident variables in truck-involved crashes. The results indicated that alcohol consumption, seatbelt use, vehicle type and accident location are the most important predictors in crash injury severity levels of truck accidents.

CART is a simple but powerful approach in data analysis and there is no predefined assumption to develop a CART model. In addition, while the correlation between explanatory and dependent variables are important in regression models, it is not a big concern in CART models (2006). Furthermore, CART models provide graphical structure including a tree with many branches and leaves for results. Graphical features assist in better understanding and interpreting the results (2011).

Despite all advantages of CART model for data analysis, instability of this model type is known as the most important disadvantage of this approach in data modelling. The ensemble models that combine two or more models to find a more robust prediction, classification or variable selection are one approach to create stable results (Dean & Wexler, 2014). Boosting and Bagging are two ensemble approaches based on DTs. Tree boosting tries to create a more accurate tree by combining many unstable and inaccurate trees. Chung (Clifton & Kreamer-Fults, 2007) used Boosted Regression Tree (BRT) to analyse single-vehicle motorcycle crashes in Taiwan. The results showed that BRT models are able to provide improved transferability than other models. Furthermore, other studies showed that boosting multiple simple trees can overcome the instability and poor accuracy of CART models (Holubowycz, 1995). Also, Bagging technique is a method for generating multiple versions of a predictor and using those to get an aggregated predictor (Pasanen & Salmivaara, 1993). Random forest is the most common bagging method that is used in some traffic safety studies to find influencing variables in traffic crashes (Appel et al, 1975, Anderson et al, 1997, Davis, 2001, Jiang et al, 2016).

Reviewing the literature, it was found that there are limited studies on pedestrian crashes at mid-blocks. Current studies mainly focus on factors influencing the pedestrian crash risk or investigate the interaction of pedestrians and drivers at mid-blocks. Those studies used limited socio-economic variables and focused on a particular age group (e.g. children). The literature review showed that there are limited studies that considered traffic volume in pedestrian crash severity. In addition, distance of pedestrian crash location to public transport stops is an important explanatory variable which was ignored in the previous studies. According to Australian Bureau of statistics, 27% of Australian population was born overseas (ABS 2013). Different culture might influence the traffic behaviour. Therefore, distance of the location pedestrian crash to public transport stops as well as originality of suburb residents as a variable which shows the influence of culture on pedestrian crash are used in this research for the first time in pedestrian crash studies.

Literature showed that DT, bagging and boosting DT have been developed in many studies. However, the accuracy of these models is not discussed in those studies. Also, it is not clear that which DT ensemble method provides more accurate results in traffic crash modelling. In this research, individual DT model and two ensemble approaches in DT are developed and comprehensively evaluated for pedestrian crash severity at mid-blocks (NOT crash injury severity). Then, the results of these three models are compared to each other to identify a more accurate model. It is the first time in traffic crash studies that three different approaches in DT are compared to introduce a more reliable method in modelling traffic crash severity.

4.3. DATA

The primary dataset used in this study is the road crash statistics (CrashStats) of Victoria, Australia. It includes data on personal characteristics (e.g. driver/ pedestrian age, gender), vehicle characteristics (e.g. vehicle type, weight), road and environment conditions (e.g. road surface, light and pavement conditions), and temporal parameters (e.g. date, day and time of the crash). In Victoria, only crashes resulting in injury to at least one of the road users involved in the accident are required to be reported to the police.

The objective of this study is to identify the factors contributing to the severity of vehicle-pedestrian crashes at mid-blocks. In crash severity analysis, each observation is one crash and its severity are related to the most severe injury sustained in the crash, as recorded in the police accident report. Note that this analysis is different from injury severity analysis where each observation represents one road user and it is possible to have multiple people involved in one crash.

In Victoria, the severity of a crash is determined by the person with the most severe injury. A fatal crash refers to a crash in which at least one person died within 30 days of a collision, while a serious injury crash refers to a crash in which at least one person was sent to the hospital (Crash Statistics Data 2016). Note that this classification is different from other schemes that use actual injury scale such as the Abbreviated Injury Scale (AIS) and may be an overestimate because some of the road users sent to hospital may only suffer minor injuries.

To investigate the variables influencing vehicle-pedestrian crash severity, data for these crashes on public roadways in the Melbourne metropolitan area from 2004 to 2013 were extracted from CrashStats for this study. Of the total of 11,625 vehicle-pedestrian crashes, 5,346 were located at mid-blocks. According to VicRoads classification, of the 5,346 vehicle-pedestrian crashes included in the study, 3.5% were fatal crashes, 45.5% were serious injury crashes, and. 51.0% were minor injury crashes. In addition to the data from CrashStats, data on the neighbourhood social and economic characteristics were extracted from the Australian Urban Research Infrastructure Network (AURIN). The AURIN database comprised the largest single resource for accessing diverse types and sources of data, spanning the physical, social, economic and ecological aspects of Australian cities, towns and communities (AURIN 2013). ArcMap GIS 10 was used to extract the social and economic variables related to each suburb where the vehicle-pedestrian collisions occurred. ArcMap GIS 10 was also used to extract the traffic volume data from the Melbourne road network database for each crash location.

Table 1 shows a summary of the categorical variables used in this study. In this study, categorical explanatory variables are grouped into 5 major groups, describing the temporal, personal, traffic and road, environment, and socio-economic characteristics. The continuous variables used in this study are shown in Table 2. The continuous variables are divided into two main groups, describing the crash location and the neighbourhood around the crash location.

Table 1. Categorical explanatory variables

Variables			Percent
Dependent variable	Crash severity	Fatal	4.9
		Serious injury	47.7
		Minor injury	47.4
Temporal	Season	Spring	23.8
		Summer	23.9
		Fall	26.2
		Winter	26.1
	Time of crash	Morning peak (7:00 - 9:00)	13.6
		Afternoon peak (16:00 – 18:00)	3.7
		Daytime off-peak (10:00 – 15:00)	35.2
		Other	47.5
	Day	Weekday	75.6
		Weekend	24.4
Personal	Pedestrian gender	Female	42.6
		Male	57.1
		Unknown	0.3
	Driver gender	Female	29.1
		Male	59.5
		Unknown	11.4
	Pedestrian age	18 and under	16.7
		19-24	14.5
		25-44	31.1
		45-64	18.1
		65-74	7.2
		75 and older	9.4
		Unknown	3.0
	Driver age	25 and under	19.3
		26-44	36.4
		45-64	24.7
		65-74	4.3
		75 and older	3.3
		Unknown	12.1

continued on following page

Table 1. Continued

Variables			Percent
Traffic and road	Vehicle type	Passenger cars	78.7
		Taxi and van	8.9
		Heavy vehicles	0.8
		Buses	1.5
		Motor and bike	4.3
		Tram and Train	3.0
		Other	2.9
	Traffic control	No control	79.8
		Stop go light and flashing	5.4
		Pedestrian light and cross walk	7.8
		Unknown	4.2
	Surface condition	Dry	83.2
		Wet, muddy, snowy and Icy	12.8
		Other	4.0
Traffic and road	Divided road type	Divided double line (DD)	9.2
		Divided single centreline representation (DS)	26.1
		Not divided (ND)	40.0
		Unknown (U)	24.7
	Speed limit	≤50 km/h	28.1
		60-70 km/h	55.5
		≥80 km/h	9.8
		Other	6.6
	Median	Yes	36.5
		No	63.5
Environment	Light conditions	Day	62.1
		Dusk/dawn	5.6
		Dark street light on	26.6
		Dark no street light	4.4
		Other	1.3
	Atmosphere conditions	Clear	85.9
		Raining and snowing	9.0
		Fog, smoke, dust and strong winds	0.6
		Other	4.6
	Land use	Residential	22.8
		Commercial	29.3
		Industrial	7.1
		Community and educational	2.7
		Sport, recreation and parks	4.9
		Not Active	33.3

continued on following page

Table 1. Continued

Variables			Percent
Social (suburb characteristics)	White collar worker [a]	< 50%	39.4
		50-80%	59.6
		> 80%	1.0
	Pink collar worker [b]	50-80%	46.5
		> 80%	53.5
	Blue collar worker [c]	> 50%	8.2
		50-80%	88.9
		> 80%	2.9
	Income	Low income (< $600)	2.3
		Middle income ($600-$2,499)	94.2
		High income (> $2,500)	3.5

a. White-collar work is performed in an office, cubicle, or other administrative setting.
b. Pink-collar work is related to customer interaction, entertainment, sales, or other service-oriented work.
c. Blue-collar work refers to manufacturing, construction, mining and agricultural businesses.

Table 2. Descriptive statistics for continuous variables

Variables			Unit	Min	Max	Mean	S.D.
Crash Location		Average daily traffic (ADT)	Vehicle per day	200	122000	13250.8	10151.4
		Average road slope	Per cent	0.0	5.0	1.3	3.0
		Number of lanes	lane	2.0	13.0	4.5	2.1
		Road width	Meter	5.0	77.0	20.1	10.8
		Lane width	Meter	2.5	5.0	3.4	0.7
		Distance from public transport stops	Meter	0.1	9800.0	134.3	392.3
Neighbourhood	Age and gender	Gender ratio (male to female)	Per cent	0.9	1.5	1.0	0.06
		Median age	Year	24.0	51.0	35.4	4.7
	Overseas Born	UK	Per cent	0.4	14.4	3.9	2.1
		Southern and eastern Europe		0.0	1.2	0.2	0.2
		Middle East		0.0	23.0	0.8	1.6
		Asia		0.0	52.4	15.7	12.9
		Indigenous persons		0.0	2.9	0.4	0.4
		Others		3.3	50.0	43.6	6.9
	Education	Degree or higher		9.2	74.5	45.6	15.7
		Certificate or diploma		14.1	73.0	34.6	13.1
	Labour force	Labour force participant rate		32.8	80.0	60.0	8.3
	Population	Suburb population (pop.) density	pop per Sq. Meter			3276.3	2506.7

4.4. METHODOLOGY

Although DT, boosted and bagged DT have been used in previous vehicle crash studies, the performance and accuracy of these models in road safety analysis have yet to be compared. This study applies two ensemble methods to examine pedestrian crashes at mid-blocks and compares the effectiveness of bagging and boosting in improving the performance of the single DT model in traffic crash analysis. In addition, repeated cross-validation (CV) is applied to individual and ensemble DT models to increase the accuracy of DT models.

Additionally, the literature review shows that identifying the relative influence of the different variables on crash severity levels has been largely neglected in most studies. In the present research, partial dependence plotting is applied for the first time in pedestrian crash analysis to show how each crash contributing factor can change the pedestrian crash severity level. Figure 1 shows the methodology of this research.

4.4.1. CART, Bagging, and Boosting

DTs can be used for classification and regression tasks. If the variable in the study is categorical, a classification tree is developed, and when a DT is used to predict a continuous variable, it is called a regression tree. In CART models, predictors at the top of the tree (parent node) are divided into several homogeneous nodes using rules. This procedure is repeated and each new node (child node) is assumed to be a parent node for the following branches.

Figure 1. Using CV in model development

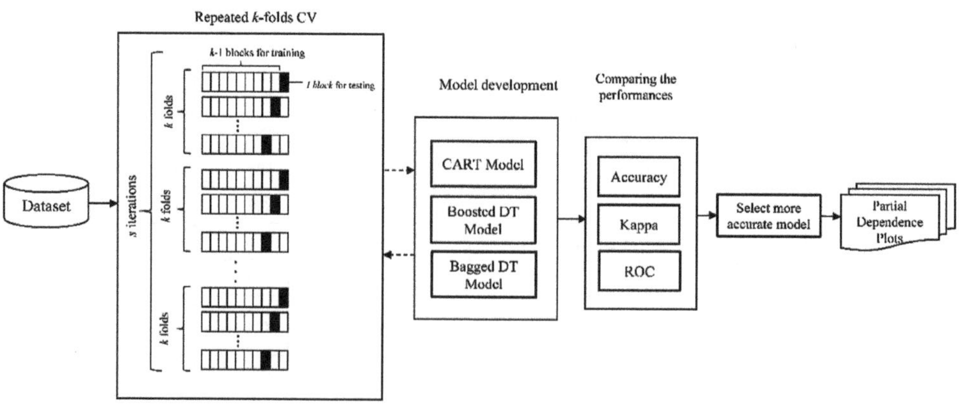

This process is continued until no further splits can be made; i.e. all child nodes are homogenous (or a user-defined minimum number of objects in the node is obtained). These final nodes are called terminal nodes or leaves, and they have no branches. Partitioning, stops when all possible threshold values for all explanatory variables (splitters) have been assessed to find the greatest improvement in the purity scores of the resultant nodes.

CART then tries to simplify the structure of the tree, which makes a smaller tree, and prevents over-fitting. The pruning process starts with the maximal tree and all branches with little impact on the predictive value of the tree are removed. CART determines the best tree by testing for error rates or costs. With sufficient data, the simplest method is to divide the sample into learning and test sub-samples. The learning sample is used to grow an overly large tree. Then the test sample is used to estimate the rate at which cases are mis-classified (e.g., adjusted by misclassification costs). The mis-classification error rate is calculated for the largest tree and also for every sub-tree. The best sub-tree is the one with the lowest or near-lowest cost, which may be a relatively small tree.

Boosting DT is an ensemble technique that tries to find a more accurate model by merging a number of trees in a sequential process. Boosting uses a forward, stage-wise procedure that only uses the results from the previous tree rather than from all other previously-fitted trees. In this approach, after the first tree is fitted, the residuals are calculated and observations with high residual values are defined as poor fit observations. In the next step, to minimise the mis-classification error rate, the estimated probabilities are adjusted by the following weights for the ith case (Equation 1):

$$w(i) = \frac{(1 + m(i)^4)}{\sum_{i=1}^{n}(1 + m(i)^4)} \tag{1}$$

where, $0 \leq m(i) \leq n$ and n is the number of fitted classification models and $m(i)$ is the number of models that mis-classified case i in the previous step. Subsequent trees are fitted to the residual of the previous tree (61). This process is repeated n times and m models adjust the estimated probabilities. Figure 2 shows the flowchart for the boosting DT method.

On the other hand, bagging or bootstrap aggregating is a different method for combining DTs or other base classifiers. Similar to boosting, the base learning algorithm is run repeatedly in a series of rounds. In each round, the base learner is trained on what is often called a "bootstrap replicate" of the original training set. Suppose the training set consists of n examples. Then a bootstrap replicate is a new training set that also consists of n examples, and which is formed by repeatedly selecting uniformly at random and with replacement m examples from the original training set (see Figure 3). This means that the same example may appear multiple times in the bootstrap replicate, or it may not appear at all.

Therefore, on each of m rounds of bagging, a bootstrap replicate is created from the original training set. A base classifier is then trained on this replicate, and the process continues. After m rounds, a final combined classifier is formed, which simply predicts with the majority vote of all of the base classifiers. While the boosting method is known as a bias reduction technique, bagging is useful to decrease the variance of models (Rezaei et al, 2013).

4.4.2. Model Development

In this research, the analysis was carried out using different packages of the statistical software R 3.2.3 (Team, 2014). CART, boosted DT and bagged DT models were developed using rpart (Therneau et al, 2010), gbm (Ridgeway, 2007), and treebag methods in the caret packages (Kuhn, 2008).

Figure 2. Boosted decision tree flowchart

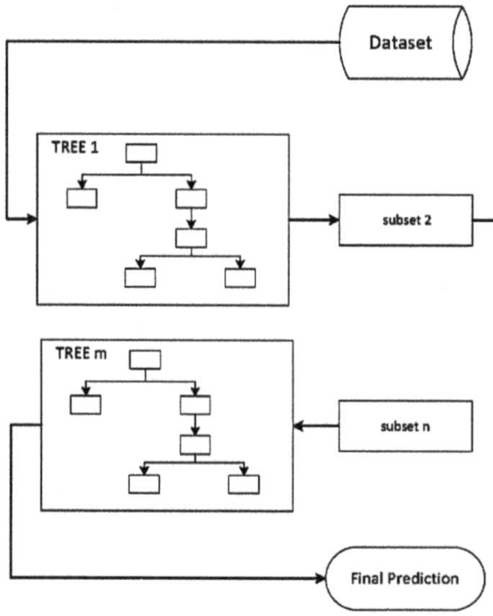

Figure 3. Bagged decision tree flowchart

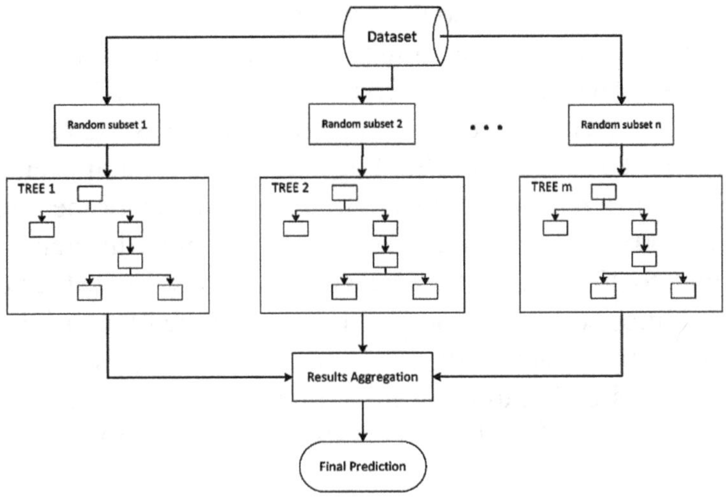

The repeated k-fold cross-validation technique was applied to develop the models, instead of dividing the data into training and testing sub-sets. K-fold cross-validation randomly divided the data into k blocks of roughly equal size. Each of the blocks was left out in turn and the other k-1 blocks were used to train the model. The left-out block was predicted and these predictions were summarized in a performance measure. This procedure was repeated s times to decrease the error and find the most robust model. The s x k estimates of performance were then averaged to obtain the overall re-sampled estimate. In this research, a 10-fold cross-validation with 5 iterations (5x10 resamples) was applied for each model and the performance of the models was estimated.

Furthermore, as mentioned before, principle behind tree growing in DT approach is to recursively partition the target variable (crash severity) to maximize "purity" in the two following child nodes. In this method, the program checks all possible input variables (splitters) as well as all possible values of the input variables to find the threshold and define a rule that leads to the greatest improvement in the purity score of the resultant nodes (Chang and Wang, 2006, Kashani and Mohaymany, 2011, Abellán et al., 2013). Therefore, in this research, program defined rules and splits the nodes using these rules to achieve maximise purity.

To find the most accurate model for each approach, the model parameters had to be optimised. The gbm package was used to optimize shrinkage, tree complexity and the number of trees. Shrinkage or learning rate was used to determine the contribution of each tree to the growing model. This parameter was used to decrease the contribution of each tree in the model. Tree complexity or interaction depth represented the depth of a tree and indicated interaction among predictor variables. Tree complexity equalled to 1 implied an additive model, while a tree with complexity 2 had 2-way interactions between variables (Friedman et al, 2001). In addition, tree complexity was used as a parameter to optimize the CART model. Finding these model parameters would be subjective and different studies used different amounts for these parameters (Elith et al, 2008, Saha et al, 2015). In this research, different interaction depths from 5 to 20, with 5 intervals, were used to optimize the boosted DT model. In addition, 0.1, 0.01 and 0.001 were assumed for shrinkage. In the CART model, 30 interaction depths (1 to 30) were analysed. Furthermore, different numbers of iterations were applied in the boosting and bagging methods (from 50 to 500 with 50 intervals) to find the most accurate models.

4.4.3. Performance Metrics

In this research, accuracy (ACC), Kappa, and area under ROC curve (AUC) were used to compare the performance of the CART, bagging DT and boosting DT models. ACC is the most widely used performance measure in machine learning methods. It is defined as the proportion of correct predictions made by the classifier relative to the size of the dataset and is presented in Equation 2.

$$ACC = \frac{r_{corr}}{r} \times 100 \quad (2)$$

where, r is the number of all possible predictions for a given problem, and r_{corr} is the number of correct predictions by the current method.

Receiver operating characteristics (ROC) curve analysis is a classical methodology that originates from signal detection theory. A ROC curve displays the relationship between sensitivity and specificity for a given classifier (Table 3). Sensitivity is a relative frequency of correctly classified positive examples (Equation 3).

$$Sens = \frac{TP}{TP + FN} = \frac{TP}{POS} \quad (3)$$

where, TP is the number of true positive examples, FN is the number of false negative examples, and POS is the number of positive examples. Specificity is the relative frequency of correctly- classified negative examples (Equation 4).

$$Spec = \frac{TN}{TN + FP} = \frac{TN}{NEG} \tag{4}$$

where, TN is the number of false negative examples, FP is the number of false positive examples, and NEG is the number of negative examples.

The horizontal axis on the ROC curve shows the false positive ratio (equivalent to 1-specificity), whereas on the vertical axis the positive ratio (sensitivity) is shown. In ROC, the classifier with the larger AUC is considered as more accurate.

Cohen's Kappa is another widely-used measure of classifiers' accuracy in machine learning techniques. Kappa is defined as follows:

$$Kappa = \frac{P_0 - P_c}{1 - P_c} \tag{5}$$

where, P_0 is the total agreement of probability, or the accuracy, and Pc is the agreement probability which is due to randomness (Ben-David, 2008).

4.4.4. Relative Importance of Variables

In addition to comparing the performance of the three DT methods, this study also examined the relative influence of predictor variables to quantify the importance of predictors on vehicle-pedestrian crashes at mid-blocks. The relative importance of each predictor in the CART and bagging DT methods is based on the number of times a variable is either selected to split a node in trees or used as a surrogate rule in the tree (Friedman et al, 2003).

In the boosting DT model, the final prediction is either a weighted average or the majority vote of all the simple classification models. Therefore, the first step is to identify the importance of each variable in every classifier, in order to find the relatively important variables in this model. For a classifier T that results in a DT with $J-1$ internal nodes, Breiman et al. (1984) proposed the following equation:

Table 3. A confusion matrix representing classification quantities for two-class problems

Correct Class	Classified as		Σ
	P (Positive Class)	N (Negative Class)	
P	TP (number of true positive examples)	FN (number of false negative examples)	POS=TP+FN (number of positive example)
N	FP (number of false positive examples)	TN (number of true negative examples)	NEG=FP+TN (number of negative example)
Σ	PP=TP+FP (number of predicted positive examples)	PN=FN+TN (number of predicted negative examples)	n= TP+FP+FN+TN

$$\Gamma_j^2(T) = \sum_{t=1}^{J-1} \xi_t^2 I(v_t = j) \quad (6)$$

where, v_t is the splitting variable associated with node t, and ξ_t^2 is the corresponding empirical improvement as a result of the split. If the predictor x_j is selected as splitting variable at node t, $I(v_t = j) = 1$, otherwise $I(v_t = j) = 0$.

The importance of predictor x_j for a collection of DTs in the boosting technique, $\{T_c\}_1^C$, is obtained by averaging or weighted averaging the importance in the set of classifiers (Breiman et al, 1984), as presented in Equation 7.

$$\Gamma_{xj}^2 = \frac{1}{C} \sum_{c=1}^{C} \Gamma_j^2(Tc) \quad (7)$$

where, C is the number of classifiers in the boosting DTs and Tc indicates the classification tree produced at the kth step.

4.4.5. Partial Dependence Plot

Visualisation of results is one of the most important advantages of DT models. The use of visual results makes the interpretation of the results of the model very simple. To visualize the results and identify the interactions between variables in boosted and bagged DT models, partial dependence plots were used in this research. Partial dependence plots depict the relationship between the response and one predictor variable, while accounting for the average effects of all other predictors (Friedman, 2001, Friedman, 2003).

4.5. RESULTS AND DISCUSSION

4.5.1. Optimizing Models

Figure 4 shows the performance of boosting DT models with different sets of shrinkage factors with fixed tree complexity. According to this figure, the model with larger shrinkage factor (0.1) would fit fewer trees with higher accuracy, whereas models with 0.01 and 0.001 shrinkage factors would fit many trees to gradually reach the maximum accuracy. Therefore, to find the most accurate models in this research the shrinkage parameter in gbm package was assumed to be 0.1.

In addition, Figures 5 and 6 illustrate the performance of boosted DT and CART models with different sets of tree complexity factors with fixed shrinkage parameters. These two figures indicate that with the same shrinkage value (0.1) relatively fewer trees are required with increasing tree complexity to fit the model. According to Figures 5 and 6, tree complexities of 20 and 12 are used in the most accurate model for the boosted DT and CART methods, respectively. In addition, different models with different numbers of resampling

Figure 4. BDT model performance with three different shrinkage factors.

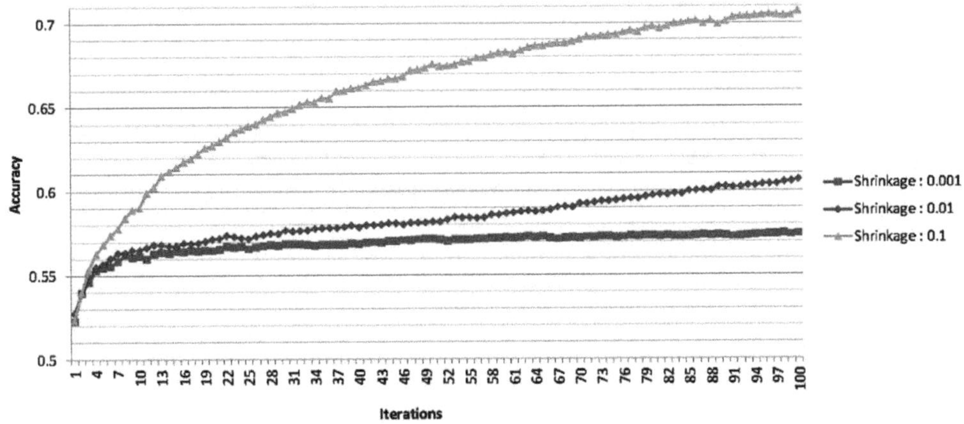

Figure 5. BDT model performance with different sets of tree complexity factors

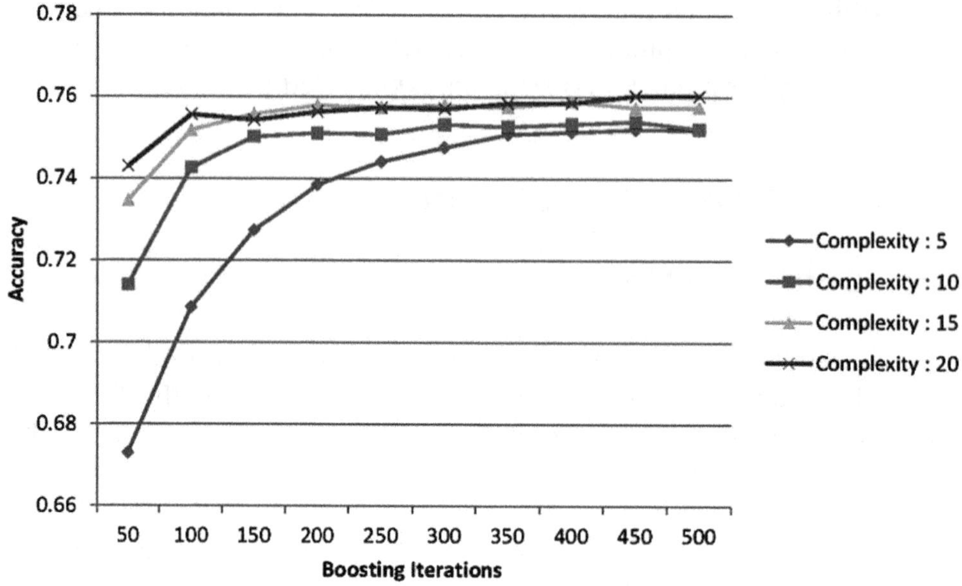

Figure 6. CART model performance with different set of tree complexity factors

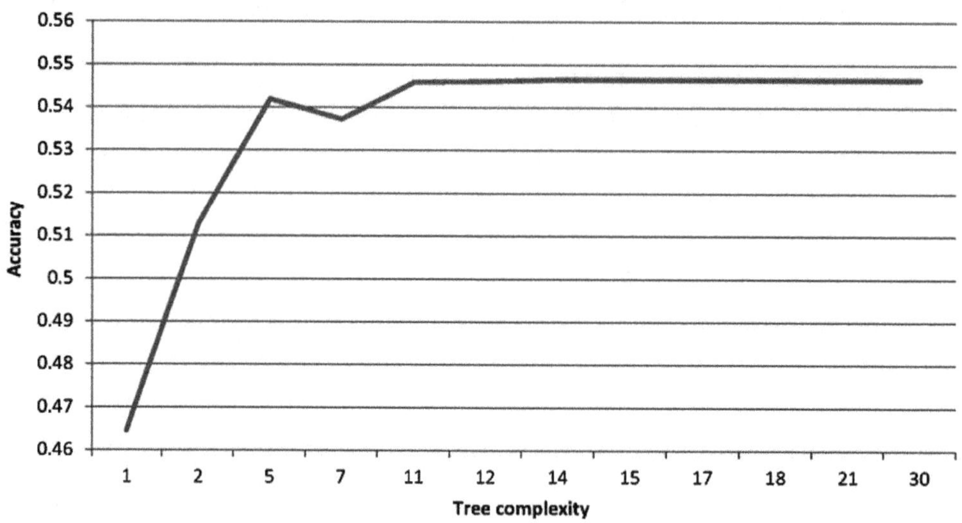

subsets were developed for bagged DT models. According to Figure 7, bagged DT with 400 resampling numbers shows the optimum accuracy.

4.5.2. Model Performance

Table 3 shows the performances of CART, boosted and bagged DTs in regard to different metrics for vehicle-pedestrian crash severity at mid-blocks. According to this table, the boosted DT and bagged DT models improve the accuracy of the CART model by 42% and 31%, respectively. In addition, the performance of ensemble tree models is better than CART models with respect to Kappa and AUC metrics. However, Table 4 reveals that boosted DT has better performance than bagged DT. According to this table, boosted DT model shows 8%, 12% and 3% more accuracy, Kappa and AUC than the bagged DT model, respectively.

Figure 7. Bagged DT model performance with different resampling subset numbers

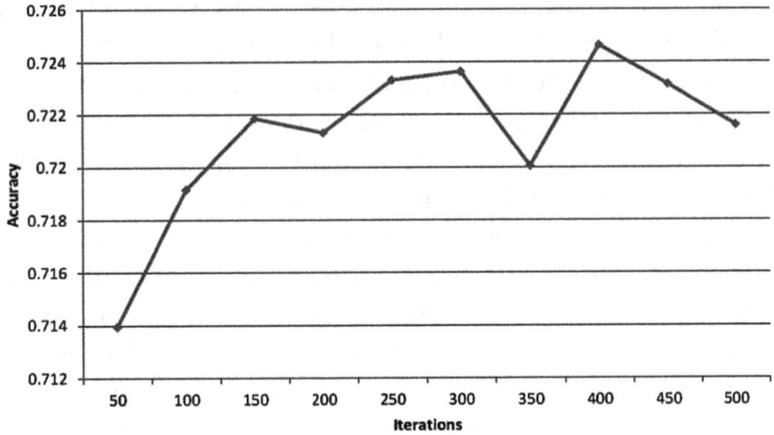

Table 4. Performance of different models

Model	Metric				
	Accuracy	Kappa	AUC	Sensitivity	Specificity
CART	0.55	0.31	0.71	0.54	0.77
Boosted DT	0.78	0.65	0.91	0.77	0.88
Bagged DT	0.72	0.58	0.88	0.72	0.85

In addition, Table 4 shows the sensitivity of the models, or the proportion of true positives. Sensitivity values in this table reflect a model's ability to correctly detect the severity level of an accident. According to these results, the ability of the boosted DT model to predict the exact severity level of an accident is higher than that of the other two models (0.77 for the boosted model compared to 0.54 for the CART model and 0.72 for the bagged DT model). Moreover, this table shows that the specificity of models (the proportion of true negatives) reflects a model's ability to predict an absence where a species does not exist. As Table 4 presents, the specificity of boosted DT is 0.88 and 14%, which is 3% higher than the bagged DT and CART models. Therefore, boosted DT is recommended over simple DT and bagged DT for analysing and modelling vehicle-pedestrian crash severity.

4.5.3. Factors Contributing to Vehicle-Pedestrian Crash Severity

Figure 8 shows the top 20 most important predictor variables for boosted DT (4.8a), bagged DT (4.8b) and CART (4.8c) models. The first and most influential factor that is identified by the three models is 'Distance to public transport'. However, there are some differences in the relative importance of the variables identified by these three models. For examples, 'Pedestrian age' and 'Speed limit' are identified as relatively more important factors in the CART model than in the bagged and boosted DT models whereas several neighbourhood social economics characteristics are identified as relatively more important factors in the bagged and boosted models than the CART model. Therefore, it is very important when using DT to choose the best method not only because of the improved accuracy but also because of the differences in the relative importance of the variables estimated.

Since the results presented in the previous section indicate that the boosted DT model has the highest accuracy in this study, the important contributing factors that have been identified by this model are highlighted and explained. Identification of these variables will assist in applying appropriate pedestrian safety counter-measures and strategies to decrease pedestrian crashes at mid-blocks. As shown in Figure (4.8a), 'Distance to public transport' and 'Average road slope' are the most important contributing factors to the severity of vehicle-pedestrian crashes, showing that these variables are a significant influencing variable in vehicle-pedestrian crash severity. The results of this study show that these factors need to be considered in vehicle-pedestrian crash studies. Whenever feasible, transportation engineers and planners should consider reducing the road gradient and the distance transit riders have to walk to access public transport. Alternatively, the speed limit could be reduced or traffic calming be installed around public transit stops to reduce the likelihood and/or severity of vehicle-pedestrian crashes.

In addition, social demographic variables, such as ethnicity, population density, gender, and educational levels of the residents in the suburb where the crash occurred are important factors contributing to the severity of vehicle-pedestrian crashes. Overall, the social-economic-demographic characteristics of the suburbs have been found to play a significant role in determining the severity of vehicle-pedestrian crashes, comprising 13 of the top 20 factors.

Figure 9 shows the partial dependence plots for the top 6 factors for the different levels of vehicle-pedestrian crash severity. In this figure, it is possible to identify the influence of different variables on vehicle-pedestrian crash severity levels. The influence of the top 6 factors on crash severity is described in the following section. Figure (9a) shows that the severity of vehicle-pedestrian crashes increases with the increase in the distance of pedestrian crash locations to public transport stops from 0 to 600m. Using different warning signals and signs around public transport stops may increase the attention given by drivers to pedestrians. More study is required to analyse vehicle-pedestrian crashes in the vicinity of public transport stops and outside this 600m distance to identify appropriate pedestrian safety programs.

Figure 8. Top 20 relative importance of predictor variables for BDT, bagged DT and CART models

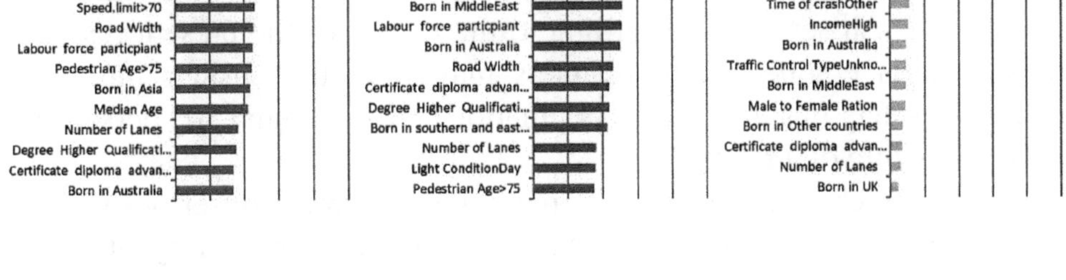

Figure (9b) shows that an increase in the average road slope to 3% or more will increase the probability of a crash being severe. This is the first time that this factor has been applied in pedestrian crash modelling. Nevertheless, the result of the present research is consistent with the results of other studies that show that the road gradient may influence the severity of other types of vehicle crashes (Dissanayake & Lu, 2002, Allen-Munley et al, 2004, Lee et al, 2008, Hosseinpour et al, 2014). Road slope (gradient) can influence drivers' sight distance and braking distance. Lack of adequate sight distance and braking distance may be the reason for the increasing severity of crashes on roads with gradients of 3% or more. Decreased speed limits on roads with more than 3% gradients could impact on driver reaction time and braking distance. Therefore, this result could be used to identify these roads and apply speed reduction strategies to improve pedestrian safety in these locations.

According to Figure 9, people with different cultures have different influences on crash severity levels. For instance, the severity of crashes increases with increased population of people born in the UK (Figure 9c) but this severity decreases in suburbs with Middle Eastern populations higher than 1% (Figure 9e). Culture may impact walking and traffic behaviour and the results of this model are consistent with the results from previous research showing that culture/family background can be an important factor influencing traffic crashes (Agran et al, 1998, Factor et al, 2007). This result will assist transportation engineers, planners and policy makers to identify the target customer segments for improving pedestrian safety. Knowing the right target audience is critical for the success of safety education and communications programs, such as publishing pedestrian safety bulletins and using warning messages on billboards.

Figure (9d) shows how population density influences vehicle-pedestrian crash severity. According to this figure, the severity of vehicle-pedestrian crashes at mid-blocks increases with increasing population density up to about 800 persons per sq. metre. Above this figure, the trend of crashes with other injuries increases, but the probability of fatal and serious injury crashes decreases. This result is similar to those of La Scala et al. (2000) and Clifton et al. (2009), who showed that increasing population density may influence pedestrian crash severity. According to the results of the present research, transportation engineers and planners may want to consider improving pedestrian safety in suburbs with population density around 800 persons per sq. metre before other suburbs.

Finally, larger traffic volumes increase the probability of serious vehicle-pedestrian crashes. According to Figure (4.9f), the probability of a fatal

Figure 9. Partial dependence plots for top 6 most important predictor variables

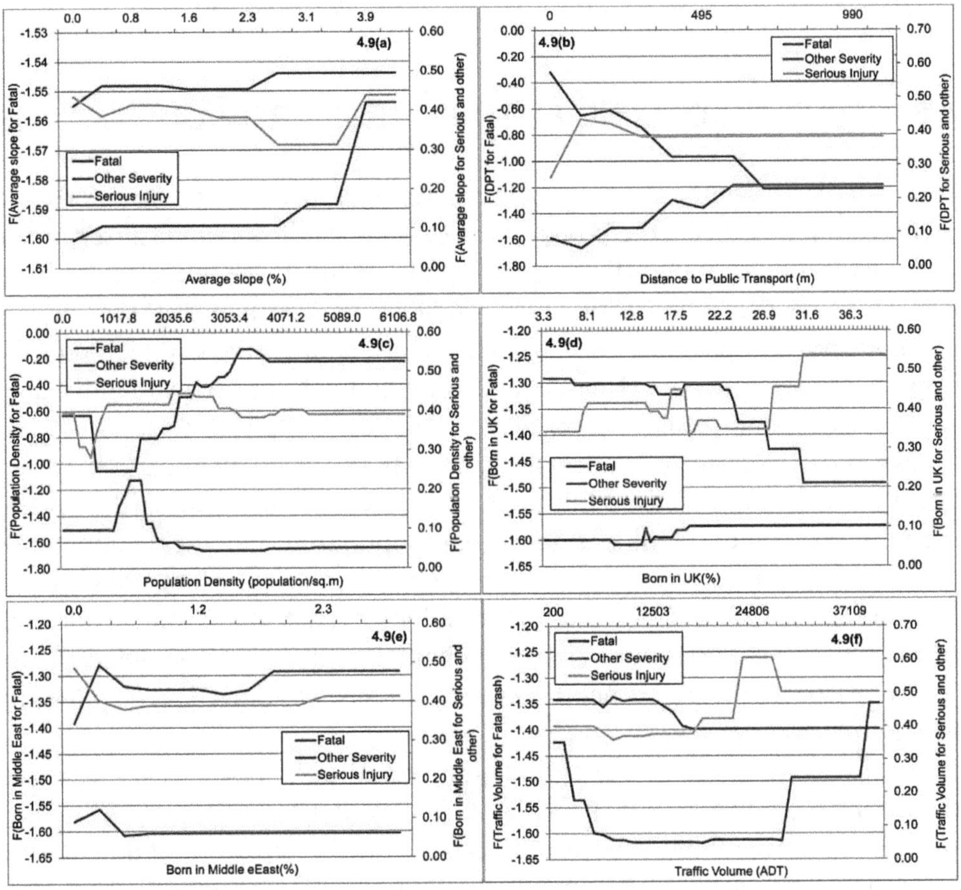

crash exhibits the classical U-shaped relationship while the probability of a minor injury crash is relatively constant in low traffic, drops significantly around 12,500-18,500 vehicles per day and then remain relatively constant again, On the other hand, the probability of a serious injury crash increases significantly when traffic volume increased beyond 20,000 vehicles per day. The results of the present research are consistent with the results of other studies that indicate that increasing traffic volume can increase pedestrian crash frequency and the probability of pedestrian crash severity (Zegeer et al, 2001, Pulugurtha & Sambhara, 2011, Morency et al, 2012). These results suggest that transportation engineers and planners may want to target roads with more than 20,000 vehicles per day to improve the safety of these vulnerable road users. More pedestrian crossings, pedestrian signals and flashing lights on these roads may assist in improving the safety of pedestrians.

4.6. CONCLUSION AND FUTURE RESEARCH DIRECTIONS

In this study, a machine learning approach was used to develop three models to predict the severity of vehicle-pedestrian crashes at mid-blocks. These models include CART, bagging DT and boosting DT. While these models have been used in some previous traffic crash studies, the accuracy of these models has not been compared previously. This study applied the CV technique to improve the accuracy of DT models, and compared the accuracy of individual DT and ensemble techniques in DT models for pedestrian crashes at mid-blocks in the Melbourne metropolitan area. In this study, vehicle-pedestrian crash severity was used as the target variable and 42 different variables, including socio-economic variables (e.g. population, income, occupation), environment variables (e.g. light conditions, land use, surface conditions), location characteristics (e.g. road slope, vehicle type, traffic volume, distance from public transport stops), personal characteristics (e.g. age, gender) and temporal variables (e.g. time, day/date of crash), were used to develop the model.

This study found that the application of CV and boosted DT improved the accuracy of DT models by 42%. This elevated the DT model accuracy from 55% to more than 75%. In addition, the results showed that the boosted DT model improved the Kappa and AUC of the CART model from 0.31 and 0.71 to 0.65 and 0.91, respectively. In addition, the boosted DT models had better performance than the bagged DT model. The accuracy of the boosted DT model was 8% and the Kappa performance was 12% better than the corresponding values for the bagged DT model. Therefore, we would recommend the use of boosted DT over the simple DT and bagged DT in analysing and modelling vehicle-pedestrian crash severity.

The results of the boosted DT model showed that the distance of pedestrian crash locations to public transport stops and road slope were the two most significant factors contributing to vehicle-pedestrian crash severity levels. In addition, this research showed that increases in the distance between pedestrian crash locations and public transport stops would increase the severity of vehicle-pedestrian crashes. Furthermore, according to this research, the probability of a vehicle-pedestrian crash being fatal or resulting in serious injury was greater on roads with average gradients of more than 3%. Other traffic and road geometry factors, such as traffic volume, speed limits, and road width ranked as the 6th, 7th, and 8th most important factors, respectively.

Interestingly, the remaining 13 of the top 20 contributing factors are related to the social-economic-demographic characteristics of the suburbs in which the vehicle-pedestrian crash occurred. According to our results, the social characteristics of the crash locations are highly significant in influencing pedestrian crash severity and more research needs to be conducted in the future to provide a better understanding of these social-spatial influences. In addition, it may be useful to develop models using the socio-economic factors related to pedestrians' or drivers' residential suburbs. Furthermore, the development of other statistical models of pedestrian crash severity at mid-blocks and comparison of the results with those for the boosted DT model may help to identify more accurate approaches to modelling this type of crash.

Moreover, one of the other advantages of DT models is that decision rules (DRs) can be extracted from their structure. These DRs can be used to identify safety problems and establish certain measures of performance. In addition, boosted regression trees can fit complex non-linear relationships, and automatically handle interaction effects between predictors. It would be worthwhile to analyse some of these interactions in future research.

REFERENCES

Abdelwahab, H., & Abdel-Aty, M. (2001). Development of artificial neural network models to predict driver injury severity in traffic accidents at signalized intersections. *Transportation Research Record: Journal of the Transportation Research Board*, 6–13.

Abellán, J., López, G., & de Oña, J. (2013). Analysis of traffic accident severity using Decision Rules via Decision Trees. *Expert Systems with Applications*, *40*(15), 6047–6054. doi:10.1016/j.eswa.2013.05.027

ABS. (2011a). *Australian Bureau of Statistics*. Canberra, Australia: Australia, Year Book.

ABS. (2011b). *Australian Bureau of Statistics*. Available: www.abs.gov.au/

ABS. (2013). *Australia's population by country of birth*. Australian Bureau of Statistics.

ABS. (2015). *National Health Survey: First Results, 2014-15*. Canberra, Australia: Australian Bureau of Statistics 2015 2015. Contract No.: No. 4364.0.55.001.

Adeli, H., & Balasubramanyam, K. V. (1988). A synergic man-machine approach to shape optimization of structures. *Computers & Structures*, *30*(3), 553–561. doi:10.1016/0045-7949(88)90289-1

Aghabayk, K., Forouzideh, N., & Young, W. (2013). Exploring a Local Linear Model Tree Approach to Car-Following. *Computer-Aided Civil and Infrastructure Engineering*, *28*(8), 581–593. doi:10.1111/mice.12011

Agran, P. F., Winn, D. G., Anderson, C. L., & Del Valle, C. (1998). Family, social, and cultural factors in pedestrian injuries among Hispanic children. *Injury Prevention*, *4*(3), 188–193. doi:10.1136/ip.4.3.188 PMID:9788088

Al-Ghamdi, A. S. (2002). Pedestrian–vehicle crashes and analytical techniques for stratified contingency tables. *Accident; Analysis and Prevention*, *34*(2), 205–214. doi:10.1016/S0001-4575(01)00015-X PMID:11829290

Anderson, R. W. G., McLean, A. J., Farmer, M. J. B., Lee, B. H., & Brooks, C. G. (1997). Vehicle travel speeds and the incidence of fatal pedestrian crashes. *Accident; Analysis and Prevention*, *29*(5), 667–674. doi:10.1016/S0001-4575(97)00036-5 PMID:9316714

Appel, H., Stuertz, G., & Gotzen, L. (1975). Influence of impact speed and vehicle parameter on injuries of children and adults in pedestrian accidents. *Proceedings of the International Research Council on the Biomechanics of Injury conference*, *3*, 83-100.

Aziz, H. M. A., Ukkusuri, S. V., & Hasan, S. (2013). Exploring the determinants of pedestrian–vehicle crash severity in New York City. *Accident; Analysis and Prevention*, *50*(0), 1298–1309. doi:10.1016/j.aap.2012.09.034 PMID:23122781

Ballesteros, M. F., Dischinger, P. C., & Langenberg, P. (2004). Pedestrian injuries and vehicle type in Maryland, 1995–1999. *Accident; Analysis and Prevention*, *36*(1), 73–81. doi:10.1016/S0001-4575(02)00129-X PMID:14572829

Ben-David, A. (2008). About the relationship between ROC curves and Cohen's kappa. *Engineering Applications of Artificial Intelligence*, *21*(6), 874–882. doi:10.1016/j.engappai.2007.09.009

Bennet, S. A., & Yiannakoulias, N. (2015). Motor-vehicle collisions involving child pedestrians at intersection and mid-block locations. *Accident; Analysis and Prevention*, *78*, 94–103. doi:10.1016/j.aap.2015.03.001 PMID:25756845

Breiman, L., Friedman, J., Stone, C. J., & Olshen, R. A. (1984). *Classification and regression trees*. New York: CRC Press.

Celikoglu, H. B. (2013). An Approach to Dynamic Classification of Traffic Flow Patterns. *Computer-Aided Civil and Infrastructure Engineering*, *28*(4), 273–288. doi:10.1111/j.1467-8667.2012.00792.x

Chang, L.-Y., & Chien, J.-T. (2013). Analysis of driver injury severity in truck-involved accidents using a non-parametric classification tree model. *Safety Science*, *51*(1), 17–22. doi:10.1016/j.ssci.2012.06.017

Chang, L.-Y., & Wang, H.-W. (2006). Analysis of traffic injury severity: An application of non-parametric classification tree techniques. *Accident; Analysis and Prevention*, *38*(5), 1019–1027. doi:10.1016/j.aap.2006.04.009 PMID:16735022

Chong, M. M., Abraham, A., & Paprzycki, M. (2005). Traffic Accident Analysis Using Machine Learning Paradigms. *Informatica (Slovenia)*, *29*(1), 89–98.

Chung, Y.-S. (2013). Factor complexity of crash occurrence: An empirical demonstration using boosted regression trees. *Accident; Analysis and Prevention*, *61*, 107–118. doi:10.1016/j.aap.2012.08.015 PMID:22975365

Clifton, K. J., & Kreamer-Fults, K. (2007). An examination of the environmental attributes associated with pedestrian–vehicular crashes near public schools. *Accident; Analysis and Prevention*, *39*(4), 708–715. doi:10.1016/j.aap.2006.11.003 PMID:17174259

Cui, Z., & Nambisan, S. (2003). Methodology for Evaluating the Safety of Midblock Pedestrian Crossings. *Transportation Research Record: Journal of the Transportation Research Board*, *1828*(1), 75-82.

Davis, G. (2001). Relating Severity of Pedestrian Injury to Impact Speed in Vehicle-Pedestrian Crashes: Simple Threshold Model. *Transportation Research Record: Journal of the Transportation Research Board*, *1773*(1), 108-13.

De Oña, J., López, G., & Abellán, J. (2013). Extracting decision rules from police accident reports through decision trees. *Accident; Analysis and Prevention*, *50*, 1151–1160. doi:10.1016/j.aap.2012.09.006 PMID:23021419

Dean, J., & Wexler, J. (2014). *What's New in SAS® Enterprise Miner™ 13.1*. Academic Press.

Dissanayake, S., & Lu, J. J. (2002). Factors influential in making an injury severity difference to older drivers involved in fixed object–passenger car crashes. *Accident; Analysis and Prevention*, *34*(5), 609–618. doi:10.1016/S0001-4575(01)00060-4 PMID:12214955

Elith, J., Leathwick, J. R., & Hastie, T. (2008). A working guide to boosted regression trees. *Journal of Animal Ecology*, *77*(4), 802–813. doi:10.1111/j.1365-2656.2008.01390.x PMID:18397250

Eluru, N., Bhat, C. R., & Hensher, D. A. (2008). A mixed generalized ordered response model for examining pedestrian and bicyclist injury severity level in traffic crashes. *Accident; Analysis and Prevention*, *40*(3), 1033–1054. doi:10.1016/j.aap.2007.11.010 PMID:18460372

Factor, R., Mahalel, D., & Yair, G. (2007). The social accident: A theoretical model and a research agenda for studying the influence of social and cultural characteristics on motor vehicle accidents. *Accident; Analysis and Prevention*, *39*(5), 914–921. doi:10.1016/j.aap.2006.12.015 PMID:17291438

Friedman, J., Hastie, T., & Tibshirani, R. (2001). The elements of statistical learning. Springer.

Friedman, J. H., & Meulman, J. J. (2003). Multiple additive regression trees with application in epidemiology. *Statistics in Medicine*, *22*(9), 1365–1381. doi:10.1002im.1501 PMID:12704603

Herabat, P., & Songchitruksa, P. (2003). A Decision Support System for Flexible Pavement Treatment Selection. *Computer-Aided Civil and Infrastructure Engineering*, *18*(2), 147–160. doi:10.1111/1467-8667.00306

Holubowycz, O. T. (1995). Age, sex, and blood alcohol concentration of killed and injured pedestrians. *Accident; Analysis and Prevention*, *27*(3), 417–422. doi:10.1016/0001-4575(94)00064-S PMID:7639925

Hung, S.-L., & Jan, J. C. (1999). Machine Learning in Engineering Analysis and Design: An Integrated Fuzzy Neural Network Learning Model. *Computer-Aided Civil and Infrastructure Engineering*, *14*(3), 207–219. doi:10.1111/0885-9507.00142

Jiang, X., Abdel-Aty, M., Hu, J., & Lee, J. (2016). Investigating macro-level hotzone identification and variable importance using big data: A random forest models approach. *Neurocomputing*, *181*, 53–63. doi:10.1016/j.neucom.2015.08.097

Kashani, A. T., & Mohaymany, A. S. (2011). Analysis of the traffic injury severity on two-lane, two-way rural roads based on classification tree models. *Safety Science*, *49*(10), 1314–1320. doi:10.1016/j.ssci.2011.04.019

Kim, J.-K., Ulfarsson, G. F., Shankar, V. N., & Kim, S. (2008). Age and pedestrian injury severity in motor-vehicle crashes: A heteroskedastic logit analysis. *Accident; Analysis and Prevention*, *40*(5), 1695–1702. doi:10.1016/j.aap.2008.06.005 PMID:18760098

Kim, J.-K., Ulfarsson, G. F., Shankar, V. N., & Mannering, F. L. (2010a). A note on modeling pedestrian-injury severity in motor-vehicle crashes with the mixed logit model. *Accident; Analysis and Prevention*, *42*(6), 1751–1758. doi:10.1016/j.aap.2010.04.016 PMID:20728626

Y.-I. Kim, S. H. Park, & S.-Y. Kho (Eds.). (2010b). *A Case Study of 'Continuous Risk Profile' Approach for Hotspots Identification on Korean Expressways. 17th ITS World Congress*, Busan, South Korea.

Koepsell, T., McCloskey, L., & Wolf, M. (2002). CRosswalk markings and the risk of pedestrian–motor vehicle collisions in older pedestrians. *Journal of the American Medical Association*, *288*(17), 2136–2143. doi:10.1001/jama.288.17.2136 PMID:12413373

Kuhn, M. (2008). Caret package. *Journal of Statistical Software*, *28*(5).

Kuhnert, P. M., Do, K.-A., & McClure, R. (2000). Combining non-parametric models with logistic regression: An application to motor vehicle injury data. *Computational Statistics & Data Analysis*, *34*(3), 371–386. doi:10.1016/S0167-9473(99)00099-7

Kwon, O. H., Rhee, W., & Yoon, Y. (2015). Application of classification algorithms for analysis of road safety risk factor dependencies. *Accident; Analysis and Prevention*, *75*, 1–15. doi:10.1016/j.aap.2014.11.005 PMID:25460086

Laflamme, L., Vaez, M., Hasselberg, M., & Kullgren, A. (2005). Car safety and social differences in traffic injuries among young adult drivers: A study of two-car injury-generating crashes in Sweden. *Safety Science*, *43*(1), 1–10. doi:10.1016/j.ssci.2004.09.001

Lee, C., & Abdel-Aty, M. (2005). Comprehensive analysis of vehicle–pedestrian crashes at intersections in Florida. *Accident; Analysis and Prevention*, *37*(4), 775–786. doi:10.1016/j.aap.2005.03.019 PMID:15869737

Lefler, D. E., & Gabler, H. C. (2004). The fatality and injury risk of light truck impacts with pedestrians in the United States. *Accident; Analysis and Prevention*, *36*(2), 295–304. doi:10.1016/S0001-4575(03)00007-1 PMID:14642884

Lord, D., van Schalkwyk, I., Chrysler, S., & Staplin, L. (2007). A strategy to reduce older driver injuries at intersections using more accommodating roundabout design practices. *Accident; Analysis and Prevention*, *39*(3), 427–432. doi:10.1016/j.aap.2006.09.011 PMID:17092474

Miles-Doan, R. (1996). Alcohol use among pedestrians and the odds of surviving an injury: Evidence from Florida law enforcement data. *Accident; Analysis and Prevention*, *28*(1), 23–31. doi:10.1016/0001-4575(95)00030-5 PMID:8924182

Miranda-Moreno, L. F., Morency, P., & El-Geneidy, A. M. (2011). The link between built environment, pedestrian activity and pedestrian–vehicle collision occurrence at signalized intersections. *Accident; Analysis and Prevention*, *43*(5), 1624–1634. doi:10.1016/j.aap.2011.02.005 PMID:21658488

Morency, P., Gauvin, L., Plante, C., Fournier, M., & Morency, C. (2012). Neighborhood Social Inequalities in Road Traffic Injuries: The Influence of Traffic Volume and Road Design. *American Journal of Public Health*, *102*(6), 1112–1119. doi:10.2105/AJPH.2011.300528 PMID:22515869

Moudon, A. V., Lin, L., Jiao, J., Hurvitz, P., & Reeves, P. (2011). The risk of pedestrian injury and fatality in collisions with motor vehicles, a social ecological study of state routes and city streets in King County, Washington. *Accident; Analysis and Prevention*, *43*(1), 11–24. doi:10.1016/j.aap.2009.12.008 PMID:21094292

Newstead, S., & D'Elia, A. (2010). Does vehicle colour influence crash risk? *Safety Science*, *48*(10), 1327–1338. doi:10.1016/j.ssci.2010.05.001

Oikawa, S., Matsui, Y., Doi, T., & Sakurai, T. (2016). Relation between vehicle travel velocity and pedestrian injury risk in different age groups for the design of a pedestrian detection system. *Safety Science*, *82*, 361–367. doi:10.1016/j.ssci.2015.10.003

Pasanen, E., & Salmivaara, H. (1993). Driving speeds and pedestrian safety in the City of Helsinki. *Traffic Engineering & Control*, *34*(6), 308–310.

Peña-garcía, A., De Oña, R., García, P., Peña-garcía, P., & de Oña, J. (2014). Effects of Daytime Running Lamps on Pedestrians Visual Reaction Time: Implications on Vehicles and Human Factors. *Procedia Engineering*, *84*, 603–607. doi:10.1016/j.proeng.2014.10.473

Pulugurtha, S. S., & Sambhara, V. R. (2011). Pedestrian crash estimation models for signalized intersections. *Accident; Analysis and Prevention*, *43*(1), 439–446. doi:10.1016/j.aap.2010.09.014 PMID:21094342

Quistberg, D. A., Howard, E. J., Ebel, B. E., Moudon, A. V., Saelens, B. E., Hurvitz, P. M., ... Rivara, F. P. (2015). Multilevel models for evaluating the risk of pedestrian–motor vehicle collisions at intersections and mid-blocks. *Accident; Analysis and Prevention*, *84*, 99–111. doi:10.1016/j.aap.2015.08.013 PMID:26339944

Reich, Y. (1997). Machine Learning Techniques for Civil Engineering Problems. *Computer-Aided Civil and Infrastructure Engineering*, *12*(4), 295–310. doi:10.1111/0885-9507.00065

Rezaei, M. A., Behzadi, G., Ahmadian, S., & Rezaei, M. (2013). Pedestrian's Accidents Prediction in Suburban Roads Using Artificial Neural Network (Case study of Amol city). *Journal of Intelligent Transportation and Urban Planning.*, *1*(1), 41–47. doi:10.18005/ITUP0101005

Ridgeway, G. (2007). Generalized Boosted Models: A guide to the gbm package. *Update.*, *1*(1), 2007.

Saha, D., Alluri, P., & Gan, A. (2015). Prioritizing Highway Safety Manual's crash prediction variables using boosted regression trees. *Accident; Analysis and Prevention*, *79*(0), 133–144. doi:10.1016/j.aap.2015.03.011 PMID:25823903

Sarkar, S., Tay, R., & Hunt, J. (2011). Logistic Regression Model of Risk of Fatality in Vehicle-Pedestrian Crashes on National Highways in Bangladesh. *Transportation Research Record: Journal of the Transportation Research Board, 2264*(1), 128-37.

Savolainen, P. T., Mannering, F. L., Lord, D., & Quddus, M. A. (2011). The statistical analysis of highway crash-injury severities: A review and assessment of methodological alternatives. *Accident; Analysis and Prevention, 43*(5), 1666–1676. doi:10.1016/j.aap.2011.03.025 PMID:21658493

Siddiqui, C., Abdel-Aty, M., & Choi, K. (2012). Macroscopic spatial analysis of pedestrian and bicycle crashes. *Accident; Analysis and Prevention, 45*, 382–391. doi:10.1016/j.aap.2011.08.003 PMID:22269522

Team, R. C. R. (2014b). *A language and environment for statistical computing*. Vienna, Austria: R Foundation for Statistical Computing.

Team, R. C. R. (2014a). A language and environment for statistical computing. R Foundation for Statistical Computing.

Thurston, D. L., & Sun, R. (1994). Machine Learning User Preferences for Structural Design. *Computer-Aided Civil and Infrastructure Engineering, 9*(3), 185–197. doi:10.1111/j.1467-8667.1994.tb00372.x

Tulu, G. S., Washington, S., Haque, M. M., & King, M. J. (2015). Investigation of pedestrian crashes on two-way two-lane rural roads in Ethiopia. *Accident; Analysis and Prevention, 78*, 118–126. doi:10.1016/j.aap.2015.02.011 PMID:25770907

Wang, C.-H., & Chen, N. (2015). A GIS-based spatial statistical approach to modeling job accessibility by transportation mode: Case study of Columbus, Ohio. *Journal of Transport Geography, 45*, 1–11. doi:10.1016/j.jtrangeo.2015.03.015

Wang, J., Zheng, Y., Li, X., Yu, C., Kodaka, K., & Li, K. (2015). Driving risk assessment using near-crash database through data mining of tree-based model. *Accident; Analysis and Prevention, 84*, 54–64. doi:10.1016/j.aap.2015.07.007 PMID:26319604

Wey, W.-M., & Chiu, Y.-H. (2013). Assessing the walkability of pedestrian environment under the transit-oriented development. *Habitat International, 38*, 106–118. doi:10.1016/j.habitatint.2012.05.004

WHO. (2013). *Pedestrian safety: a road safety manual for decision-makers and practitioners*. World Health Organization.

Wier, M., Weintraub, J., Humphreys, E. H., Seto, E., & Bhatia, R. (2009). An area-level model of vehicle-pedestrian injury collisions with implications for land use and transportation planning. *Accident; Analysis and Prevention*, *41*(1), 137–145. doi:10.1016/j.aap.2008.10.001 PMID:19114148

Yingying, Z., Danya, Y., Qiu, T. Z., Lihui, P., & Yi, Z. (2012). Pedestrian Safety Analysis in Mixed Traffic Conditions Using Video Data. Intelligent Transportation Systems. *IEEE Transactions on.*, *13*(4), 1832–1844.

Yu, R., & Abdel-Aty, M. (2014). Analyzing crash injury severity for a mountainous freeway incorporating real-time traffic and weather data. *Safety Science*, *63*, 50–56. doi:10.1016/j.ssci.2013.10.012

Zajac, S. S., & Ivan, J. N. (2003). Factors influencing injury severity of motor vehicle–crossing pedestrian crashes in rural Connecticut. *Accident; Analysis and Prevention*, *35*(3), 369–379. doi:10.1016/S0001-4575(02)00013-1 PMID:12643954

Zegeer, C., Stewart, J., Huang, H., & Lagerwey, P. (2001). Safety effects of marked versus unmarked crosswalks at uncontrolled locations: Analysis of pedestrian crashes in 30 cities. *Transportation Research Record: Journal of the Transportation Research Board*, 56–68.

Zheng, Y., Chase, T., Elefteriadou, L., Schroeder, B., & Sisiopiku, V. P. (2015). Modeling vehicle–pedestrian interactions outside of crosswalks. *Simulation Modelling Practice and Theory*, *59*, 89–101. doi:10.1016/j.simpat.2015.08.005

Chapter 5
Neighbourhood Influences on Vehicle-Pedestrian Crash Severity

ABSTRACT

Socioeconomic factors are known to be contributing factors to vehicle-pedestrian crashes. Although several studies have examined the socioeconomic factors related to the locations of crashes, few studies have considered the socioeconomic factors of the neighbourhoods where road users live in vehicle-pedestrian crash modelling. In vehicle-pedestrian crashes in the Melbourne metropolitan area, 20% of pedestrians, 11% of drivers, and only 6% of both drivers and pedestrians had the same postcode for the crash and residency locations. Therefore, an examination of the influence of socioeconomic factors of their neighbourhoods, and their relative importance will contribute to advancing knowledge in the field, as very limited research has been conducted on the influence of socioeconomic factors of both the neighbourhoods where crashes occur and where pedestrians live. In this chapter, neighbourhood factors associated with road users' residents and location of crash are investigated using BDT model. Furthermore, partial dependence plots are applied to illustrate the interactions between these factors. The authors found that socioeconomic factors account for 60% of the 20 top contributing factors to vehicle-pedestrian crashes. This research reveals that socioeconomic factors of the neighbourhoods where road users live and where crashes occur are important in determining the severity of crashes, with the former having a greater influence. Hence, road safety counter-measures, especially those focussing on road users, should be targeted at these high-risk neighbourhoods.

DOI: 10.4018/978-1-5225-7943-4.ch005

5.1. INTRODUCTION

Relatively few studies, however, had examined the contribution of socioeconomic factors, such as culture, income and level of education, on vehicle-pedestrian crashes. Campos-Outcalt et al. (2002) examined the influence of race and ethnicity on pedestrian crashes in Arizona, and revealed that the rates and circumstances of pedestrian deaths were affected by these factors. In addition, several studies had examined the influences of income and education level on vehicle-pedestrian crashes (Dougherty et al, 1990, Borrell et al, 2005, Lyons et al, 2008, Cottrill et al, 2010).

In general, two main approaches were used to examine the influence of socioeconomic variables on vehicle-pedestrian crashes. Some studies used the socioeconomic characteristics of the neighbourhood where pedestrians lived (Borrell, 2005) while other studies used the socioeconomic factors related to the neighbourhoods where the crash occurred (Amoh-Gyimah et al, 2016, Toran Pour et al, 2017). However, limited or no study has investigated the socioeconomic factors related to both types of neighbourhoods or examined their relative importance.

Studies show that social and economic factors related to location of crashes could influence on vehicle-pedestrian severity level. For instance, Wier et al. (2009) and Graham et al. (2005) showed that the proportion of low-income households and the proportion of people without access to a motor vehicle were contributing factors for vehicle-pedestrian crash injury severity. Therefore, understanding the social and economic factors related to location of crashes may assist road safety professionals to target suburbs to apply site-specific pedestrian safety programs and improving vehicle-pedestrian safety issue in these suburbs. Meanwhile, other studies showed that socio-economic factors could influence road users' behaviour (Wilde, 1976, Ishaque et al, 2008). For instance, studies showed that ethnicity and family background were important factors associated with traffic crashes (Coughenour, 2017). Therefore, using drivers' and pedestrians' residency neighbourhood social and economic factors can assist in identifying target suburbs to apply different road user behaviour change programs and improve traffic safety knowledge of road users in these suburbs.

This research aims to examine the influence of socioeconomic factors of the neighbourhoods where the crashes occur, the neighbourhoods where drivers live, and the neighbourhoods where pedestrians live, as well as examining their relative importance. It will examine the neighbourhoods where the road users live (residency neighbourhood) and where the crashes occur (crash neighbourhood) on vehicle-pedestrian crash severity, while controlling for the influences of roadway, road user, vehicle and environmental factors. It will contribute to advancing knowledge in the field as very limited research has been conducted to examine the influence of socioeconomic factors of both the neighbourhoods where the crashes occur and where the pedestrians live.

5.2. DATA AND METHODOLOGY

Data on vehicle-pedestrian crashes on public roads that occurred at mid-blocks in the Melbourne metropolitan area from 2004 to 2013 were extracted from the Victorian Road Crash Information database (Crash Statistics Data 2016). The summary of the data is presented in Table 1. In addition, data on the socioeconomic factors were extracted from ABS (2013).

Since information on the postcodes of crash locations and the addresses of the persons involved in the crashes (residency location) are available in the crash database, socioeconomic data at the postcode level were extracted from the ABS. GIS was then used to merge the crash information and the socioeconomic data. The final dataset used contained 3,577 crashes, of which 152 (4%) were fatal crashes, 1,679 (47%) were serious injury crashes and 1,746 (49%) were other injury crashes. It should be noted that 20% of pedestrians, 11% of drivers and only 6% of both drivers and pedestrians had the same postcode for the crash location and their residency location (Toran Pour, Moridpour, Tay, & Rajabifard, 2017).

All data including crash data, social and economic data, and traffic data were added as separate layers in ArcMap GIS and merged using the postcodes of the crash locations and the postcodes of the road users' addresses. The final dataset included data on traffic and road characteristics, personal characteristics, and socioeconomic data for the locations of crashes and road users' residency neighbourhoods. As reported in Chapter 4, the boosted DT (BDT) model showed a better performance than DT models. Therefore, the final dataset was applied in the BDT model to identify factors contributing to vehicle-pedestrian crashes (the BDT model technique is described in Chapter 4). Figure 1 shows the method applied in this research.

In the present research, fitted BDT models were obtained using the "gbm" library (Ridgeway, 2007) in the R software (Team, 2014) in the caret package (Kuhn, 2008). To develop the BDT model, the repeated k-fold cross-validation technique was applied. The dataset was randomly divided into k blocks of roughly equal size, instead of dividing the data into training and testing subsets. In each iteration, one block was omitted and the other k-1 blocks were used to train the model. Each k block was omitted once and the omitted block was used for prediction. These predictions are summarized in a performance measure (e.g. accuracy). This procedure was repeated s times to decrease the error and find the most robust model. The (s x k) estimates of performance were then averaged to obtain the overall re-sampled estimate.

In this research, a 10-fold cross-validation with 5 iterations (Toran Pour et al, 2017) was applied to each model and the performances of the models were estimated. In addition to the error rate, interaction depth and shrinkage were used to evaluate the BDT models. The shrinkage or learning rate was used to determine the contribution of each tree to the growing model. This parameter was used to decrease the contribution of each tree in the model. Tree complexity or interaction depth represents the depth of a tree and shows the interaction among predictor variables. Based on previous research, the interaction depth and shrinkage parameters were assumed to be 15 and 0.1

Table 1. Distribution of categorical variables

Factors		Severity		
		Fatal	**Serious Injury**	**Other Injury**
Season	Autumn	5.6%	49.5%	44.9%
	Spring	5.0%	53.3%	41.7%
	Summer	5.3%	49.9%	44.8%
	Winter	6.3%	48.6%	45.1%
Time	1Peak (7:00-9:00)	4.7%	40.4%	54.9%
	2Peak (16:00-19:00)	15.8%	61.8%	22.4%
	Off-Peak (9:00-16:00)	3.2%	46.4%	50.4%
	Other	6.8%	55.1%	38.1%
Day	Weekdays	5.1%	48.7%	46.2%
	Weekend	7.0%	55.2%	37.8%
Pedestrian Age	19-24	3.2%	52.0%	44.8%
	25-44	5.5%	51.6%	42.9%
	45-64	4.9%	47.7%	47.4%
	65-74	9.3%	47.5%	43.2%
	75+	13.0%	51.6%	35.4%
	Other	0	50.0%	50.0%
	Under 18	2.1%	49.5%	48.4%

Factors		Severity		
		Fatal	Serious Injury	Other Injury
Pedestrian Gender	Female	3.6%	47.8%	48.6%
	Male	7.1%	52.0%	40.9%
	Unknown		66.7%	33.3%
Driver Age	26-44	5.4%	49.1%	45.5%
	45-64	4.8%	48.4%	46.8%
	65-74	5.0%	54.4%	40.6%
	75+	6.6%	48.7%	44.7%
	Other	9.1%	63.6%	27.3%
	Under 25	6.9%	53.8%	39.3%
Driver Gender	Female	4.2%	49.7%	46.0%
	Male	6.3%	50.6%	43.2%
	Unknown	0	25.0%	75.0%
Pedestrian Gender	Female	3.6%	47.8%	48.6%
	Male	7.1%	52.0%	40.9%
	Unknown	0	66.7%	33.3%
Vehicle Type	Bus	0	55.6%	44.4%
	Heavy Vehicle	27.3%	27.3%	45.4%
	Motorcycle and Bicycle	2.5%	54.3%	43.2%
	Other	22.5%	46.9%	30.6%
	Passenger car	5.2%	51.2%	43.6%
	Taxi and van	4.9%	39.9%	55.2%
	Tram and train	7.1%	51.8%	41.1%
Traffic Condition	Flash	3.4%	50.4%	46.2%
	No control	5.7%	51.0%	43.3%
	P-Crossing	5.5%	48.6%	45.9%
	P-light	6.6%	50.0%	43.4%
	School crossing	16.7%	16.5%	66.7%
	Unknown	4.3%	42.7%	53.0%
Surface Condition	Dry	5.3%	49.7%	45.0%
	Icy	0	100.0%	0
	Other	3.4%	25.4%	71.2%
	Wet	7.8%	59.3%	32.9%
Atmosphere Condition	Clear	5.7%	50.0%	44.3%
	Dust	0	0	100.0%
	Fog	0	66.7%	33.3%
	Raining	5.4%	62.3%	32.3%
	Strong wind	0	50.0%	50.0%
	Unknown	3.2%	23.8%	73.0%

Factors		Severity		
		Fatal	Serious Injury	Other Injury
Light Condition	Dark lights on	9.4%	60.0%	30.6%
	Dark no lights	20.0%	48.2%	31.8%
	Day	3.2%	46.1%	50.7%
	Dusk/Dawn	4.9%	52.9%	42.2%
	Other	0	54.5%	45.5%
Speed Limit	<60	1.1%	44.0%	54.9%
	>70	16.5%	58.9%	24.6%
	60-70	6.1%	53.2%	40.7%
	Other	1.0%	33.6%	65.4%
Land Use	Commercial	3.3%	45.2%	51.5%
	Community and Education	3.7%	61.1%	35.2%
	Industrial	10.7%	52.6%	36.7%
	Residential	5.2%	55.5%	39.3%
	Sport and recreation	8.3%	46.9%	44.8%
	Undefined	6.2%	49.9%	43.9%
Road dividing type	Divided double line (DD)	12.8%	54.4%	32.8%
	Divided single centreline (DS)	6.3%	53.7%	40.0%
	Not divided (ND)	4.2%	49.1%	46.7%
	Unknown (U)	3.8%	46.8%	49.4%
Existence of road median	Yes	5.5%	50.2%	44.3%
	No	1.0%	40.0%	59.0%

Figure 1. Methodology applied in this research

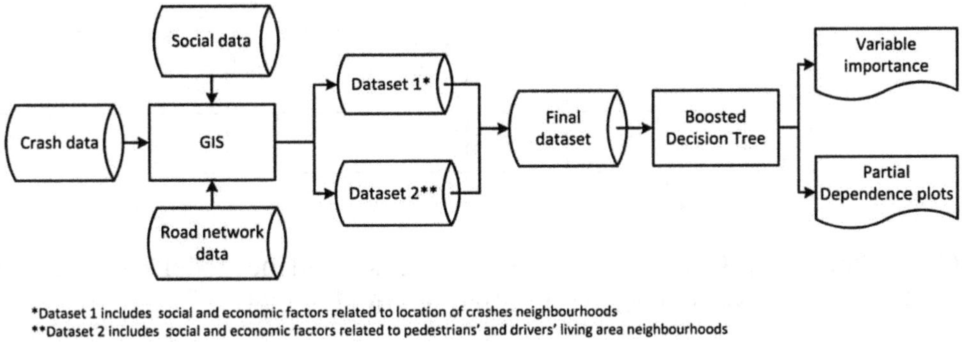

Table 2. Descriptive statistics for continuous variables

Factors		Unit	Mean	Std. Deviation
Traffic and Geometry	Traffic volume	Vehicles per day	13,376.63	10,201.44
	Average gradient	Per cent	1.37	3.01
	Distance to public transport station	Metre	137.73	422.62
	Road width	Metre	20.03	10.93
Social and Economical	Population density	Pop. per sq. kilometre	2,794.43	2,058.17
	Indigenous	Per cent	.40	.30
	Median age	Per cent	35.53	4.85
	Median income	Per cent	635.99	192.92
	Born in Australia	Per cent	57.07	15.30
	Born UK	Per cent	3.83	1.96
	Born in Asia	Per cent	12.45	11.01
	Born in India	Per cent	3.01	2.29
	Born in Middle East	Per cent	.59	1.64
	Born in SE Europe	Per cent	2.39	2.24
	Born in other countries	Per cent	13.66	4.17
	Born not stated	Per cent	6.99	4.38

Factors		Unit	Mean	Std. Deviation
Social and Economical	Secondary and under	Per cent	16.55	7.26
	Technical and Further	Per cent	7.29	1.88
	University or other Tertiary Institution	Per cent	23.35	14.39
	Other Type of Education	Per cent	28.67	10.33
	Employed rate	Per cent	93.70	2.80
	Labour Force Participation	Per cent	61.36	7.17
	White Collar job	Per cent	20.45	5.97
	Blue Collar job	Per cent	21.80	9.03
	Pink Collar job	Per cent	57.46	5.29
	Use train to commute	Per cent	6.61	4.20
	Use bus to commute	Per cent	1.41	1.32
	Use tram to commute	Per cent	5.16	7.12
	Use other type of transport	Per cent	3.98	2.70
	Use private car	Per cent	55.63	18.11
	Use walk to commute	Per cent	7.45	11.53

(Toran Pour et al, 2017), respectively, in this study and the model was repeated 2,000 times (boosting iterations) to find the final model.

5.3. RESULTS AND DISCUSSION

Identifying factors which may have important relationships with vehicle-pedestrian crash severity could assist traffic safety professionals in choosing the most appropriate safety countermeasures, such as road geometry modifications, improving traffic safety knowledge or changing road users' behaviours through social marketing campaigns. The top 20 factors identified by the BDT models are shown in Figure 2. To facilitate comparison, the scores indicating the importance of each factor were scaled so that the top factor has a score of 100. In general, the 20 most important factors can be grouped into five categories: roadway and traffic, road user characteristics, drivers' residency neighbourhood, pedestrians' residency neighbourhood and crash neighbourhood factors. In addition to the importance plot, the partial

Figure 2. Top 20 most important variables in BDT model for vehicle-pedestrian crashes

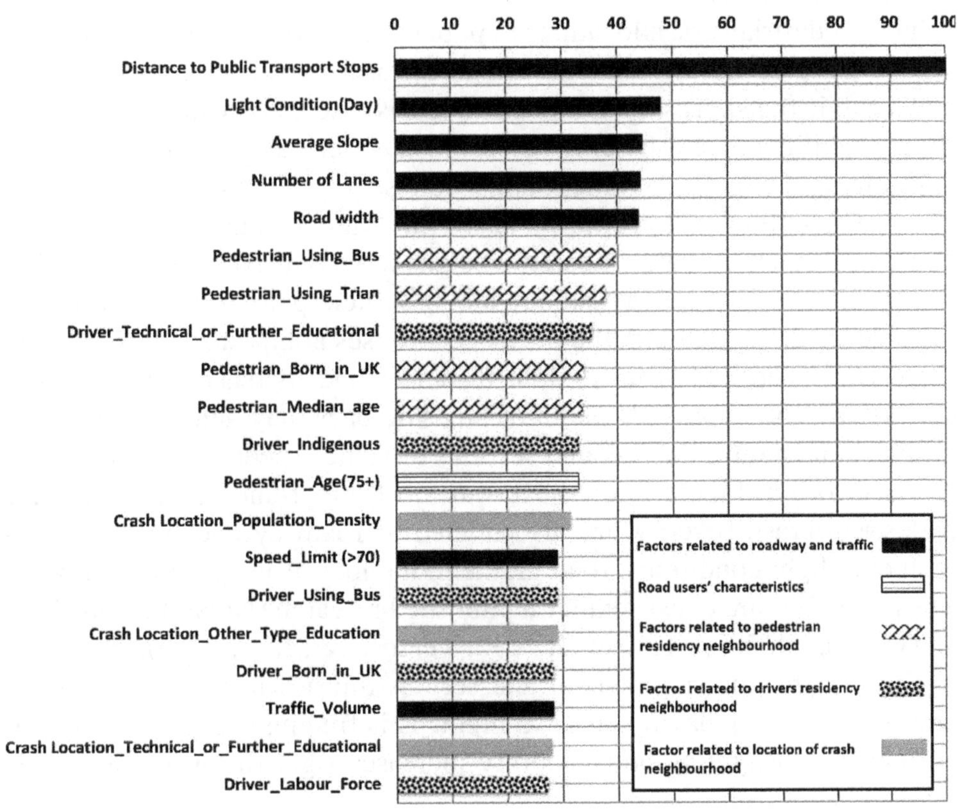

dependence plots for these factors were also generated, and these are shown in Figures 5.3 to 5.7, to facilitate the interpretation of the results.

As shown in Figure 2, seven of the top 20 factors are roadway- and traffic-related factors, one is related to the characteristics of road users involved in crashes, four are socioeconomic factors of the pedestrians' residency neighbourhood, five are related to drivers' residency neighbourhood, and three are related to the socioeconomic characteristics of the crash location. In addition, six of the socioeconomic characteristics of the drivers' and pedestrians' residency neighbourhoods are ranked ahead of all the socioeconomic factors related to the crash location. These findings suggest that the socioeconomic factors related to drivers' and pedestrian' residency neighbourhoods are more important than the socioeconomic factors related to the locations of crashes.

5.3.1. Roadway-and-Traffic-Related Factors

According to Figure 2, seven of the top 20 factors are roadway- and traffic-related factors, and five of them are ranked at the top. Overall, the most important contributing factor identified by the BDT model is the distance of vehicle-pedestrian crash locations to public transport stops. Furthermore, as shown in Figure 5.3(a), an increase in the distance between the crash location and a public transport stop up to 600 meters is associated with an increase in the probability of a crash being severe (fatal or serious injury). This finding is similar to the results of other published research (Toran Pour et al, 2017). Therefore, the application of lower speed limits around public transport stops or the use of on-site safety posters or signs to warn drivers and pedestrians to be more careful might assist in reducing vehicle-pedestrian crash severity in these areas. In addition, the presence of buses at bus stops may decrease the drivers' and pedestrians' sight distances and increase the probability of crashes. The improved design of bus stop bays may help to resolve this problem and reduce the risk and severity of vehicle-pedestrian crashes.

As shown in Figure 2, the next four most important factors in vehicle-pedestrian crash severity are also related to roadway and traffic factors, including light conditions, road gradient, number of lanes and road width. These findings are consistent with published research (Noland & Oh, 2004, Harwood et al, 2008, Tulu et al, 2015, Verzosa & Miles, 2016). According to Figure 3(b), the risk of fatal and serious injuries in vehicle-pedestrian crashes is lower in daytime than at night. This finding is expected, because an increase in lighting and visibility increases sight distance. Hence, the

Figure 3. Roadway- and-traffic-related factors, (a) Distance from public transport stops, (b) Light conditions, (c) Road gradient, (d) Number of lanes

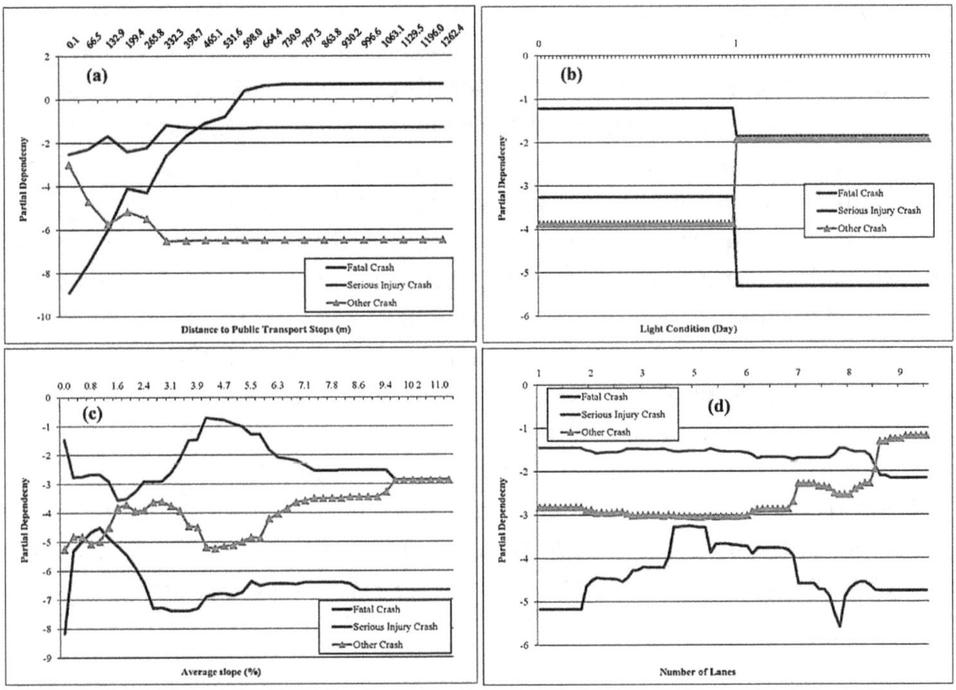

installation of street lighting in vehicle-pedestrian crash hotspots would alleviate this safety issue.

In addition, according to Figure 3(c), increasing the road gradient up to about 1% increases the likelihood of a crash being fatal, while the risk of serious injury increases between 1% and 4% road gradient, and further increases in road gradient increase the likelihood of a crash resulting in only minor injury. The road gradient affects the breaking distance and driver sight distance. Therefore, the application of lower speed limits on roads with gradients and the provision of sufficient warning to drivers to reduce their speed could reduce vehicle-pedestrian crash severity.

Furthermore, according to Figure 3(d), increasing the number of lanes increases the severity of vehicle-pedestrian crashes. On wider roads, pedestrians need more time to cross. Therefore, the installation of pedestrian crossing facilities and road medians on wider roads to provide a safe refuge when crossing could be effective means to improve pedestrian safety.

5.3.2. Road Users' Characteristics

Four road users' characteristics were examined in this study because these were the only characteristics for which data were available. These characteristics included the age (six groups) and gender (three groups) of the drivers, and the age (seven groups) and gender (three groups) of the pedestrians involved in the crashes. Interestingly, as shown in Figure 2, only one of these 19 variables were ranked in the top 20 contributing factors. For example, driver gender (male) had an importance score of only 2.4 and was not ranked in the top 100 variables. Among other reasons, the age and gender of road users are often used to capture their attitudes and behaviours, and these characteristics have been found to be significant in many studies (Sze & Wong, 2007, Tay et al, 2011, Rifaat et al, 2012, Rifaat et al, 2017). However, some of these influences may have been captured by the socioeconomic characteristics of the drivers' and pedestrians' residency neighbourhoods.

Pedestrians aged 75 and older is the only variable to be ranked in the top 20 factors affecting vehicle-pedestrian crash severity. In addition, according to Figure 4, pedestrians over 75 years of age experience an increased risk of

Figure 4. Road users' characteristics (Population over 75 years of age)

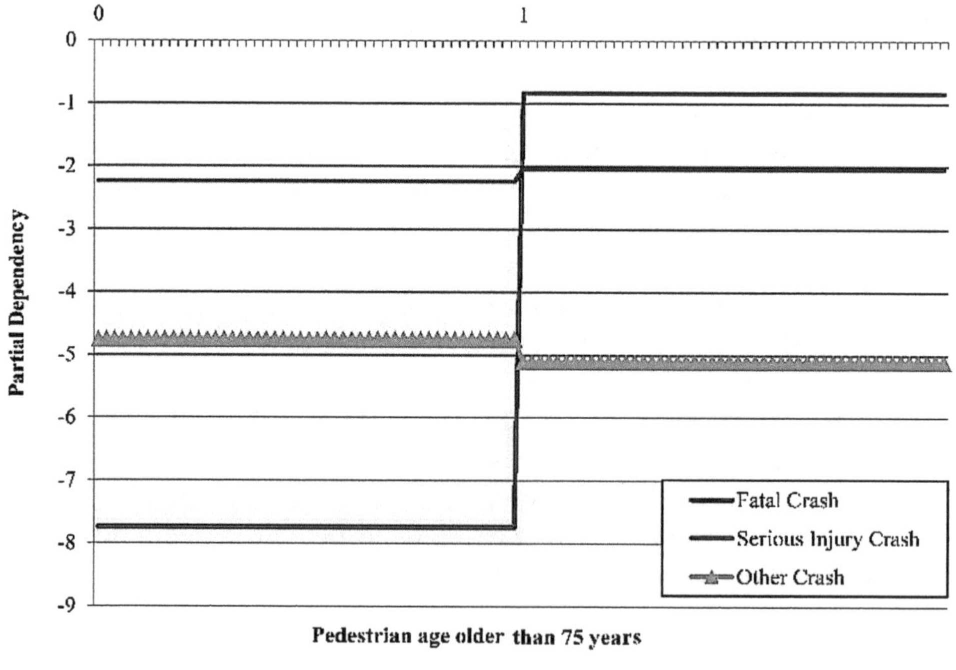

suffering fatal or serious injury in a vehicle-pedestrian collision. This result was expected, because of the increased fragility of older pedestrians. This result is similar to those of other studies showing that an increase in the age of pedestrians increases the probability of serious and fatal crashes (Sze & Wong, 2007).

5.3.3. Pedestrians' Residency Neighbourhood Characteristics

As shown in Figure 5(a), variations in the percentage of public transport usage for commuting in pedestrians' residency neighbourhoods have a considerable influence on vehicle-pedestrian crash severity. According to the figure, increases in the percentage of public transport usage in pedestrians' residency neighbourhoods are associated with increases in the probability of fatal and serious injury crashes, although the risk of fatal injury appears to plateau around 1.5%. Moreover, as discussed previously, the probability of

Figure 5. Pedestrian residency neighbourhood factors, (a) Bus commutes, (b) Suburb median age, (c) Born in UK

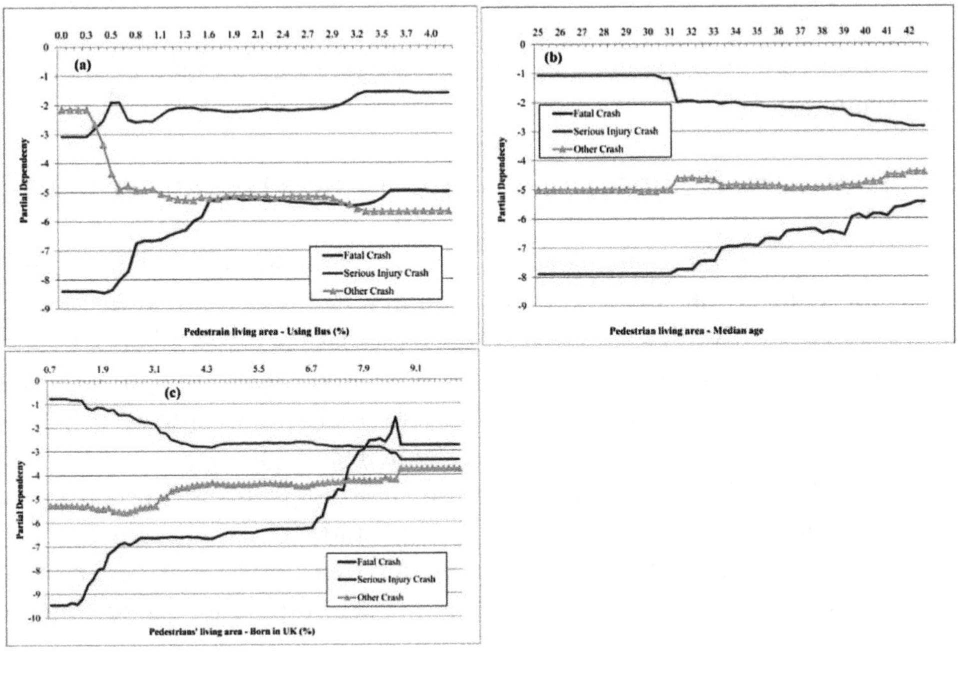

a crash being serious or fatal increases with increasing distance from public transport stops. These results are expected, because people need to walk to access to public transport stops, and pedestrians often jaywalk to catch the bus or tram. Therefore, the probability of a vehicle-pedestrian crash being severe is higher for pedestrians who live in suburbs with high percentages of bus usage.

As shown in Figure 2, the median age of people living in pedestrians' residency neighbourhoods is an important factor contributing to vehicle-pedestrian crash severity. Furthermore, as shown in Figure 5(b), the probability of a vehicle-pedestrian crash resulting in a fatality or serious injury is higher for pedestrians who live in suburbs with a median age higher than 30, and this risk increases with increasing median age. Therefore, pedestrian- related road safety campaigns or health promotion activities should target these suburbs.

According to Figure 2, country of birth is another contributing factor in vehicle-pedestrian crash severity. People from different family backgrounds and cultures may have different attitudes and walking behaviours, which could have an impact on vehicle-pedestrian crash severity. Previous research has shown that ethnicity and family background are important factors associated with traffic crashes (Coughenour et al, 2017). Our results, as highlighted in Figure 5(c), suggest that suburbs with higher proportions of people born in the UK could be targeted for pedestrian safety educational programs or campaigns. These programs could increase traffic safety knowledge, especially safe walking knowledge, and improve pedestrian safety for people living in these suburbs.

5.3.4. Drivers' Residency Neighbourhood Characteristics

As shown in Figure 2, the level of education in the drivers' residency neighbourhoods is an important contributory factor in vehicle-pedestrian crash severity. According to Figure 6(a), an increase in the number of people with technical and further education in the drivers' residency neighbourhoods increases the risk of a serious injury crash instead of a minor injury crash. This result is consistent with previous studies which have found that drivers with higher levels of education accept more risk in driving and are more involved in traffic crashes (Shinar et al, 2001, Hassan et al, 2017). Therefore, driver safety education, pedestrian awareness campaigns, and other behavioural change programs should target neighbourhoods with higher proportions of people with technical and further education.

Figure 6. Drivers' residency neighbourhood factors, (a) Tertiary education, (b) Indigenous people, (c) Bus commutes, (d) Born in UK

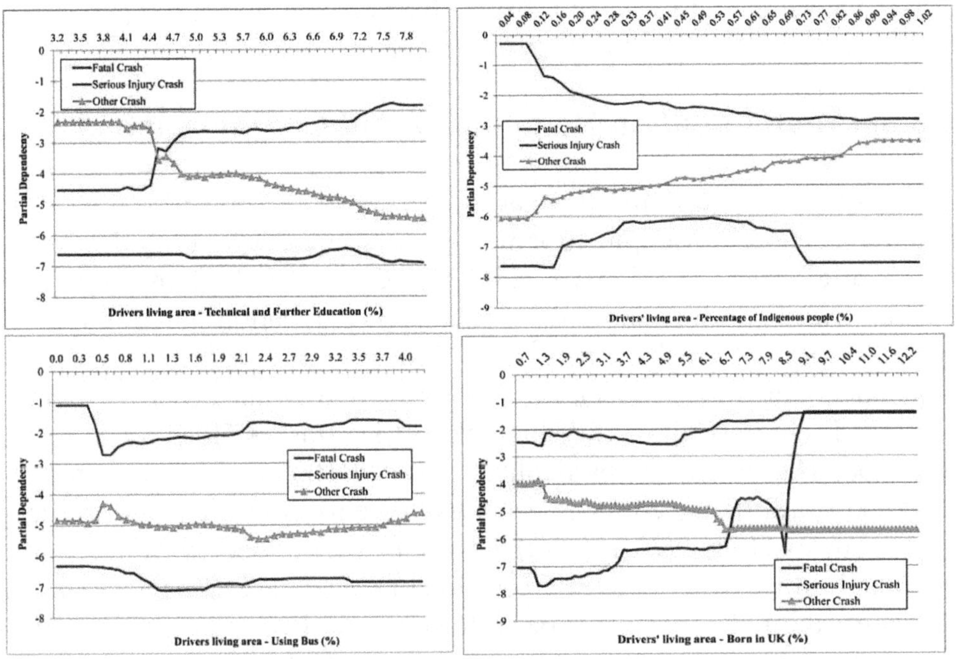

As shown in Figure 2, the ethnicity of the people living in the drivers' residency neighbourhoods is an important factor contributing to vehicle-pedestrian crash severity. According to Figure 6(b), an increase in the percentage of people living in drivers' residency neighbourhoods with Indigenous backgrounds from 0.2 to 0.6 increases the probability of fatal vehicle-pedestrian crashes but decreases the probability of vehicle-pedestrian crashes resulting in serious injuries.

Furthermore, as shown in Figure 2, the percentage of public transport use in drivers' residency neighbourhoods influences the severity of vehicle-pedestrian crashes. In addition, according to Figure 6(c), the severity of crashes is generally lower in suburbs where more people use public transport. Drivers living in these suburbs may be more aware of and have better attitudes to pedestrians and public transport users. Therefore, education programs and campaigns to inform drivers about pedestrian safety should be targeted at neighbourhoods with lower percentages of public transport use.

Moreover, as shown in Figure 6(d), increases in the percentage of people living in drivers' residency neighbourhoods who were born in the UK are associated with increases in vehicle-pedestrian crash severity levels. Therefore, driver safety education and other behavioural change programs should be targeted at neighbourhoods with around 0.6 per cent Indigenous people and higher proportions of people born in the UK.

5.3.5. Crash Location Neighbourhood Characteristics

As shown in Figure 2, the population density of the crash location is an important factor contributing to vehicle-pedestrian crash severity. This result is similar to those of La Scala et al. (2000), Clifton et al. (2009) and Toran Pour et al. (2017), which showed that changes in population density influence vehicle-pedestrian crash severity. According to Figure 7(a), increases in the population density are associated with slight and gradual increases in the likelihood of serious injuries in vehicle-pedestrian crashes. However, increases in the population density, particularly between 1000 and 2000 people per square km, are associated with decreases in the likelihood of fatal crashes.

As shown in Figure 7(b) and (c), larger numbers of people with technical or "other" education living in the location of crashes are associated with an increase in the probability of serious vehicle-pedestrian crashes. These results were expected, because the land use and jobs available in each suburb attract people with specific education types and levels, incomes and occupations. In addition, these factors may influence driving and walking behaviours in the neighbourhood, the quality of the road, and the swiftness of emergency response in the event of a crash.

Therefore, site-specific road safety countermeasures, such as road safety audits, engineering treatments and enforcement activities should target neighbourhoods with high population densities and higher percentages of people with technical or "other" education. In addition, site-specific road safety messages could be used to alert road users and change driver and pedestrian behaviours in these areas.

5.4. CONCLUSION

The identification of factors contributing to vehicle-pedestrian crash severity could assist transportation engineers, road safety professionals and policy makers in developing and implementing effective countermeasures to reduce the number of pedestrian deaths and injuries. In this research,

Figure 7. Crash neighbourhood factors, (a) Population density, (b) Other type of education, (c) Tertiary education

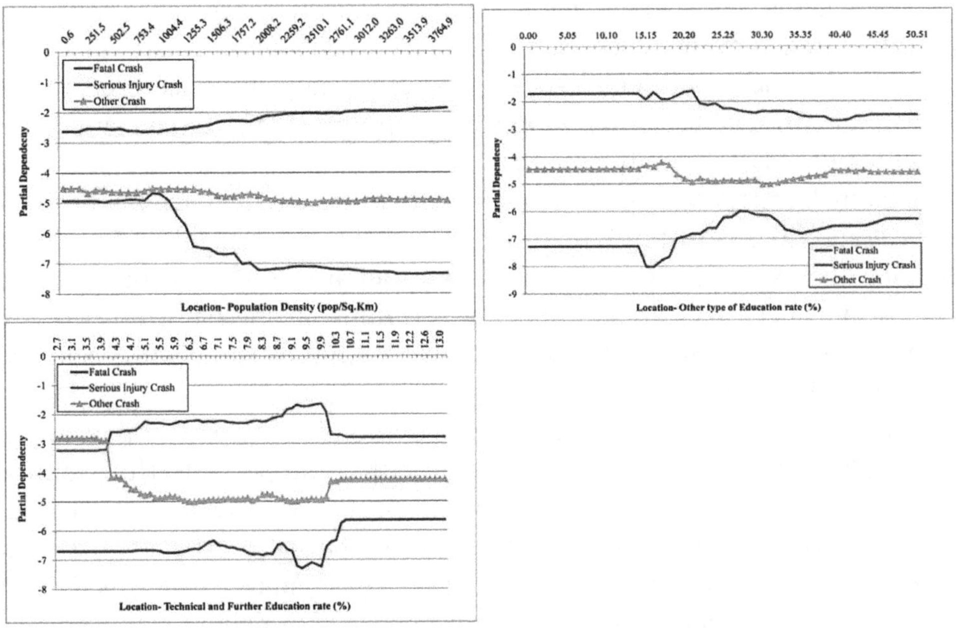

the BDT model was applied to identify the contribution of socioeconomic factors related to locations of crashes, and also pedestrians' and drivers' residency neighbourhoods. The results of this research provide valuable information to assist road safety professionals in targeting the most appropriate neighbourhoods to implement different safety measures related to pedestrians and drivers, as well as planning site-specific safety measures to reduce vehicle-pedestrian crashes.

This study has found that neighbourhood socioeconomic characteristics account for 12 out of the 20 most important variables in vehicle-pedestrian crash severity at mid-blocks. Moreover, this research reveals that nine of these 12 socioeconomic variables are related to pedestrians' and drivers' residency neighbourhoods, which shows the importance of factors related to residency neighbourhoods compared to factors related to the location of crashes.

This research has found that public transport use and family background are the two most important factors affecting vehicle-pedestrian crash severity that are related to pedestrians' residency neighbourhoods. According to the results of this research, neighbourhoods with high public transport usage, and higher proportions of people born in the UK, should be targeted for measures to improve the safety of pedestrians, such as pedestrian safety education programs.

This study has also found that the level of education, ethnicity, and usage of public transport in the drivers' residency neighbourhoods are important contributory factors to vehicle-pedestrian crash severity. This research has shown that drivers' behaviour modification programs need to be targeted at neighbourhoods with higher proportions of people with technical and trade education, higher proportions of people born in the UK, and neighbourhoods with around 0.6 percent of the population with Indigenous backgrounds. Raising the awareness of drivers in these neighbourhoods about vehicle-pedestrian safety and the development and implementation of other driver safety campaigns will decrease the number of injuries and deaths related to vehicle-pedestrian crashes.

Further, this research has found that population density, and the percentage of people with technical or trade education have important effects on the location and severity of vehicle-pedestrian crashes. This research provides evidence to support the recommendation that site-specific engineering, enforcement and safety messages should be applied in neighbourhoods with higher population densities and larger proportions of people with technical or trade education.

REFERENCES

ABS. (2011a). *Australian Bureau of Statistics*. Canberra, Australia: Australia, Year Book.

ABS. (2011b). *Australian Bureau of Statistics*. Available: www.abs.gov.au/

ABS. (2013). *Australia's population by country of birth Canberra*. Australian Bureau of Statistics.

ABS. (2015). *National Health Survey: First Results, 2014-15*. Canberra, Australia: Australian Bureau of Statistics 2015. Contract No.: No. 4364.0.55.001.

Amoh-Gyimah, R., Sarvi, M., & Saberi, M. (Eds.). (2016). Investigating the Effects of Traffic, Socioeconomic, and Land Use Characteristics on Pedestrian and Bicycle Crashes: A Case Study of Melbourne, Australia. *Transportation Research Board 95th Annual Meeting*.

Borrell, C., Plasència, A., Huisman, M., Costa, G., Kunst, A., & Andersen, O. (2005). Education level inequalities and transportation injury mortality in the middle aged and elderly in European settings. *Injury Prevention, 11*(3), 138–142. doi:10.1136/ip.2004.006346 PMID:15933403

Campos-Outcalt, D., Bay, C., Dellapenna, A., & Cota, M. K. (2002). Pedestrian fatalities by race/ethnicity in Arizona, 1990–1996. *American Journal of Preventive Medicine, 23*(2), 129–135. doi:10.1016/S0749-3797(02)00465-8 PMID:12121801

Clifton, K. J., Burnier, C. V., & Akar, G. (2009). Severity of injury resulting from pedestrian–vehicle crashes: What can we learn from examining the built environment? *Transportation Research Part D, Transport and Environment, 14*(6), 425–436. doi:10.1016/j.trd.2009.01.001

Cottrill, C. D., & Thakuriah, P. (2010). Evaluating pedestrian crashes in areas with high low-income or minority populations. *Accident; Analysis and Prevention, 42*(6), 1718–1728. doi:10.1016/j.aap.2010.04.012 PMID:20728622

Coughenour, C., Clark, S., Singh, A., Claw, E., Abelar, J., & Huebner, J. (2017). Examining racial bias as a potential factor in pedestrian crashes. *Accident; Analysis and Prevention, 98*, 96–100. doi:10.1016/j.aap.2016.09.031 PMID:27716495

Dougherty, G., Pless, I. B., & Wilkins, R. (1990). Social class and the occurrence of traffic injuries and deaths in urban children. *Canadian Journal of Public Health, 81*(3), 204–209. PMID:2361207

Graham, D., Glaister, S., & Anderson, R. (2005). The effects of area deprivation on the incidence of child and adult pedestrian casualties in England. *Accident; Analysis and Prevention, 37*(1), 125–135. doi:10.1016/j.aap.2004.07.002 PMID:15607283

Harwood, D. W., Bauer, K. M., Richard, K. R., Gilmore, D. K., Graham, J. L., Potts, I. B., & … . (2008). *Pedestrian Safety Prediction Methodology. NCHRP Web-only Document 129: Phase III*. Washington, DC: Transportation Research Board.

Hassan, H. M., Shawky, M., Kishta, M., Garib, A. M., & Al-Harthei, H. A. (2017). Investigation of drivers' behavior towards speeds using crash data and self-reported questionnaire. *Accident; Analysis and Prevention, 98*, 348–358. doi:10.1016/j.aap.2016.10.027 PMID:27837722

Ishaque, M. M., & Noland, R. B. (2008). Behavioural Issues in Pedestrian Speed Choice and Street Crossing Behaviour: A Review. *Transport Reviews*, *28*(1), 61–85. doi:10.1080/01441640701365239

Kuhn, M. (2008). Caret package. *Journal of Statistical Software*, *28*(5).

LaScala, E. A., Gerber, D., & Gruenewald, P. J. (2000). Demographic and environmental correlates of pedestrian injury collisions: A spatial analysis. *Accident; Analysis and Prevention*, *32*(5), 651–658. doi:10.1016/S0001-4575(99)00100-1 PMID:10908137

Lyons, R. A., Towner, E., Christie, N., Kendrick, D., Jones, S. J., Hayes, M., ... Phillips, C. (2008). The Advocacy in Action Study a cluster randomized controlled trial to reduce pedestrian injuries in deprived communities. *Injury Prevention*, *14*(2), e1. doi:10.1136/ip.2007.017632 PMID:18388222

Noland, R. B., & Oh, L. (2004). The effect of infrastructure and demographic change on traffic-related fatalities and crashes: A case study of Illinois county-level data. *Accident; Analysis and Prevention*, *36*(4), 525–532. doi:10.1016/S0001-4575(03)00058-7 PMID:15094404

Ridgeway, G. (2007). Generalized Boosted Models: A guide to the gbm package. *Update.*, *1*(1), 2007.

Rifaat, S. M., Tay, R., & de Barros, A. (2012). Urban Street Pattern and Pedestrian Traffic Safety. *Journal of Urban Design*, *17*(3), 337–352. doi:10.1080/13574809.2012.683398

Rifaat, S.M., Tay, R., Raihan, S.M., Fahim, A., & Touhidduzzaman, S.M. (2017). Vehicle-Pedestrian crashes at Intersections in Dhaka city. *The Open Transportation Journal, 11*(1).

Shinar, D., Schechtman, E., & Compton, R. (2001). Self-reports of safe driving behaviors in relationship to sex, age, education and income in the US adult driving population. *Accident; Analysis and Prevention*, *33*(1), 111–116. doi:10.1016/S0001-4575(00)00021-X PMID:11189114

Sze, N. N., & Wong, S. C. (2007). Diagnostic analysis of the logistic model for pedestrian injury severity in traffic crashes. *Accident; Analysis and Prevention*, *39*(6), 1267–1278. doi:10.1016/j.aap.2007.03.017 PMID:17920851

Team, R. C. R. (2014). A language and environment for statistical computing. R Foundation for Statistical Computing.

Toran Pour, A., Moridpour, S., Tay, R., & Rajabifard, A. (2017). Modelling pedestrian crash severity at mid-blocks. Transportmetrica A. *Transportation Science*, *13*(3), 273–297.

Toran Pour, A., Moridpour, S., Tay, R., & Rajabifard, A. (2017). Neighbourhood Influences on Vehicle-Pedestrian Crash Severity. *Journal of Urban Health*, *94*(6), 855-868.

Tulu, G. S., Washington, S., Haque, M. M., & King, M. J. (2015). Investigation of pedestrian crashes on two-way two-lane rural roads in Ethiopia. *Accident; Analysis and Prevention*, *78*, 118–126. doi:10.1016/j.aap.2015.02.011 PMID:25770907

Verzosa, N., & Miles, R. (2016). Severity of road crashes involving pedestrians in Metro Manila, Philippines. *Accident; Analysis and Prevention*, *94*, 216–226. doi:10.1016/j.aap.2016.06.006 PMID:27340839

VICROADS. (2015). *Principal Bicycle Network (PBN)*. Vicoria, Australia: Victorian Government Data Directory.

Wier, M., Weintraub, J., Humphreys, E. H., Seto, E., & Bhatia, R. (2009). An area-level model of vehicle-pedestrian injury collisions with implications for land use and transportation planning. *Accident; Analysis and Prevention*, *41*(1), 137–145. doi:10.1016/j.aap.2008.10.001 PMID:19114148

Wilde, G. J. S. (1976). Social Interaction Patterns in Driver Behavior: An Introductory Review. *Human Factors*, *18*(5), 477–492. doi:10.1177/001872087601800506

Chapter 6
Spatial and Temporal Distribution of Pedestrian Crashes

ABSTRACT

In order to develop effective and targeted safety programs, the location and time-specific influences on vehicle-pedestrian crashes must be assessed. Therefore, spatial autocorrelation was applied to the examination of vehicle-pedestrian crashes in geographic information systems (GISs) to identify any dependency between time and location of these crashes. Spider plotting and kernel density estimation (KDE) were then used to determine the temporal and spatial patterns of vehicle-pedestrian crashes for different age groups and gender types. Temporal analysis shows that pedestrian age has a significant influence on the temporal distribution of vehicle-pedestrian crashes. Furthermore, men and women have different crash patterns. In addition, the results of the spatial analysis show that areas with high risk of vehicle-pedestrian crashes can vary during different times of the day for different age groups and gender types. For example, for the age group between 18 and 65, most vehicle-pedestrian crashes occur in the central business district (CBD) during the day, but between 7:00 pm and 6:00 am, crashes for this age group occur mostly around hotels, clubs, and bars. Therefore, specific safety measures should be implemented during times of high crash risk at different locations for different age groups and gender types, in order to increase the effectiveness of the countermeasures in preventing and reducing the vehicle-pedestrian crashes.

DOI: 10.4018/978-1-5225-7943-4.ch006

6.1. INTRODUCTION

Pedestrians are known as vulnerable road users in road safety literature because they are more likely to be harmed or injured in traffic crashes. Pedestrians are about four times more likely to be injured in traffic crashes than other road users (Elvik, 2009). In addition, because their body is exposed and unprotected in traffic crashes, they are 23 times more likely to be killed than vehicle occupants (Miranda-Moreno et al, 2011). According to the World Health Organisation's report, every year about 1.24 million people are killed in traffic crashes in the world and more than 22% of these deaths are pedestrians (WHO 2013).

In Australia, vehicle-pedestrian crashes account for more than 13% of total fatal crashes. Every year, pedestrians are involved in about 1,100 traffic crashes in Melbourne and about 38 pedestrians are killed in these traffic crashes, which comprise about 18% of total pedestrian fatalities in Australia (Pink, 2010). Therefore, pedestrians and other vulnerable road users are specifically targeted in the recent Road Safety Agenda of the Victorian government (VicRoads, 2015). Design and implementation of effective countermeasures to improve the safety of these vulnerable road users will require not only a better understanding of the major crash contributing factors but the temporal-spatial patterns of vehicle-pedestrian crashes as well.

Spatial and temporal characteristics of traffic crashes are known to be important factors in traffic crash in many countries. For instance, different studies show that spatial and temporal parameters have an influence on traffic crash, including vehicle-pedestrian crashes in different states of U.S. (Levine et al, 1995, Aguero-Valverde & Jovanis, 2006, Li et al, 2007). In addition, a report from the National Highway Traffic Safety Administration (NHTSA) shows that location and time of crashes are main influencing factors on vehicle-pedestrian crashes in U.S. (Nhtsa, 2015).

Several studies have shown that these variables are also significant in traffic crashes in other countries. For instance, Al-Shammari et al. (2009) show that time and location of crashes are two important variables in vehicle-pedestrian crashes in the Kingdom of Saudi Arabia. Fox et al. (2015), Hosseinpour et al. (2013), and Loo et al. (2005) show the importance of location and time of crash in vehicle-pedestrian crashes in Colombia, Malaysia and Hong Kong, respectively.

Pedestrian age and gender type could influence on walking behaviour. For instance, females spend more time than men walking in their local environments and walking increased with age (2010). Therefore, many studies tried to identify the influence of pedestrian age and gender types on vehicle-pedestrian crashes. These studies identified age and gender types as two contributing factors in vehicle-pedestrian crashes (Al-Ghamdi, 2002, Henary et al, 2006, Holland & Hill, 2007). These studies revealed that age and gender types could influence on frequency and severity of vehicle-pedestrian crashes. Therefore, these two factors could influence on spatial and temporal distribution of vehicle-pedestrian crashes.

This chapter aims to identify the temporal and spatial distribution of vehicle-pedestrian crashes for different pedestrians' age groups and gender types. Specifically, it aims to answer the following research questions:

RQ1: Is there any spatial dependency between pedestrian age and gender and the crash location?
RQ2: What time-of-day are hot times for each age group and gender type?
RQ3: Do crash hot spots vary with time-of-day?
RQ4: Where are the crash hot spots for each age and gender group?

To answer these questions, spatial autocorrelation is applied in Geographical Information System (GIS) to examine vehicle-pedestrian crashes to identify any dependency between time and location of crashes for different pedestrians' age groups and gender types. Spider plots and Kernel Density Estimation (KDE) are then used to determine the temporal and spatial patterns of vehicle-pedestrian crashes at different time periods and for different pedestrians' age groups and gender types.

In the next section of the paper, the literature on vehicle-pedestrian crashes will be reviewed, with a focus on temporal and spatial analysis. In section three, an introduction to spatial autocorrelation and KDE are presented together with a description of the data and methodology used in this research. The results of this research will be presented afterwards. The final section of the paper will provide a summary of the outcomes and presents directions for future research.

6.2. LITERATURE REVIEW

Literature review shows that age and gender type could influence on walking behaviour (Bentley et al, 2010, Gómez et al, 2010, Van Dyck et al, 2010, Sundquist et al, 2011). For instance, Bentley et al. (2010) identified that female and elder people are more active in local environment. Therefore, many studies identified age and gender as two contributing factors in vehicle-pedestrian crashes (Al-Ghamdi, 2002, Henary et al, 2006, Holland et al, 2007). Tay et al. (2011) revealed that elder and female pedestrians were more influenced in vehicle-pedestrian crashes.

The review of the literature found many studies that had conducted spatial and temporal analyses of motor vehicle crashes. Black (1991) applied temporal, spatial and spatial-temporal autocorrelation analysis techniques to examine highway collisions on Indiana toll roads in U.S. He applied von Neumann's ratio, Moran's I, nearest-neighbour analysis, and a spatial-temporal autocorrelation coefficient to show the applicability of these techniques in temporal and spatial collision analysis. In another study, Levine et al. (1995) examined spatial patterns in motor vehicle crashes in Honolulu, U.S. They used GIS analysis to describe the spatial distribution of crash locations in their study area. In addition, Andrey and Yagar (1993) conducted a temporal analysis to examine the collision risks during and after rain events in Calgary and Edmonton in Canada. They applied a matched sample approach to examine the crash data between 1979 and 1983.

Aguero and Jovanis (2006) applied full Bayes hierarchical models with spatial and temporal effects and space-time interactions to examine injury and fatal crashes in Pennsylvania, U.S. They found spatial correlation in their crash data and that correlation was more important in road segment and intersection level crash models. In another study, Li et al. (2007) used a GIS-based Bayesian approach to analyse the spatial-temporal patterns of motor vehicle crashes in Houston, U.S. They found the spatial-temporal analysis method to be useful in identifying and ranking roadway segments with high risk of vehicle crashes.

In another study, Al-Shammari et al. (2009) showed the contributing factors in vehicle-pedestrian crashes in Riyadh. They showed that risk of vehicle-pedestrian crashes is more on Wednesdays. Also, this study showed that this risk is higher between 4:00pm and 12:00am than other period of time.

Plug et al. (2011) used spatial, temporal and spatiotemporal techniques in GIS to study single vehicle crash patterns in Western Australia. In this study, they used visualisation techniques, such as KDE and different types of plots. Their results showed that there were significant differences in spatial and temporal patterns of single vehicle crashes.

Moreover, the review of published research on pedestrian crashes found many studies that focused only on the spatial pattern of pedestrian crashes. In these studies, different statistical models were developed to identify the spatial variables that influence pedestrian crashes (Schneider et al, 2004). For instance, Siddiqui et al. (2012) applied a Bayesian spatial technique to model pedestrian and bicycle crashes in traffic analysis zones and found spatial correlations between pedestrian and bicycle crashes. In other studies, KDE was applied to identify pedestrian crash patterns and hot spots (Schneider et al, 2004, Truong et al, 2011). Moreover, Hosseinpour et al.(2013), applied 4 different numeric models to identify influence of road characteristics on vehicle-pedestrian crashes in Malaysia. In this study they found location of crashes could be a contributing factor in this type of crash.

The spatial temporal analyses conducted thus far had mainly examined motor vehicle crashes as a whole and did not focus on vulnerable road users while the pedestrian studies focused only on the spatial distribution of crashes. Loo et al. (2005) applied nearest neighbourhood analysis in GIS to show that the vehicle-pedestrian crashes are clustered in Hong Kong Commercial and Business Districts (CBD). Furthermore, they showed the distribution of vehicle-pedestrian crashes during the day and week and identified vehicle-pedestrian crash hotspots in this area. Also, Blasquez and Celis (2013) applied this approach to examine vehicle crashes involving child pedestrians in Santiago, Chile. In this study, they applied KDE to identify the critical areas for child pedestrian safety. They then applied Moran's Index to identify the correlation between spatial and other variables for those crashes. Furthermore, Fox et al. (2015) applied Bayesian maximum entropy (BME) method to explore influence of location of vehicle-pedestrian crashes and showed pedestrian crash fatality hotspots in Cali, Colombia. In this study mean of the spatial covariance range obtained as a function of the 3 spatial aggregations using the 9-month temporal aggregation and mean of the temporal covariance range obtained as a function of the 3 temporal aggregations using the section spatial aggregation.

In summary, the review of published literature revealed that there are limited studies focusing on both temporal and spatial analyses of motor vehicle crashes, and fewer that focussed on pedestrians and other vulnerable road users. Since vehicle-pedestrian crashes would have significantly different crash characteristics from vehicle-vehicle crashes, studies focusing on vehicle-pedestrian crashes would provide useful insights to improve the safety of these vulnerable road users. Also, literature shows many studies had identified age and gender types as two contributing factors to vehicle-pedestrian crashes. However, limited studies had explored influence of these factors on the temporal and spatial distribution of vehicle-pedestrian crashes.

6.3. DATA AND METHODOLOGY

Data for all vehicle-pedestrian crashes on public roadways in the Melbourne metropolitan area from 2004 to 2013 were extracted from the Victoria interactive crash statistics application (Crash Statistics Data 2016). In Victoria, only crashes resulting in deaths or injuries are legally required to be reported to the police. A total of 12,279 vehicle-pedestrian crashes were recorded over the 10-year period. After removing incomplete and invalid data, 9,826 pedestrian crashes were used to show the influence of pedestrians' age and gender on the temporal and spatial distributions of vehicle-pedestrian crashes. Table 1 shows a summary of the data used in this research.

Figure 1 shows the methodology of this research. The global Moran's Index is used to show the spatial autocorrelation between locations and pedestrians' age and gender. Pedestrian age is categorised into four groups: below 18 years old, 18 to 34, 35 to 65, and more than 65 years of age. Next, different spider plots were applied to identify hot times for vehicle-pedestrian crashes for different age groups and gender types. Moreover, KDE was applied to explore areas with high risk of vehicle-pedestrian crashes for each pedestrian age and gender group (Toran Pour, Moridpour, Tay, & Rajabifard, 2017).

6.3.1 Spatial Autocorrelation

The Global Moran's I measures the spatial autocorrelation between two features, which in this study, are the location and time of vehicle-pedestrian crashes. For a set of features and their associated attributes, the Global Moran's I value ranges from -1 (indicating perfect dispersion or random) to $+1$ (perfect correlation). This index can be calculated using Equation 1.

Table 1. Summary of variables

Pedestrian Gender			Age Groups				Total
			>18	18-34	35-64	Over 65	
Female	Time Categories	12 am-6:59 am	34	161	56	47	298
		7 am-8:59 am	141	226	187	245	799
		9 am-14:59 pm	302	414	392	734	1842
		3pm-6:59pm	244	389	241	271	1145
		7pm-11:59pm	109	283	140	99	631
	Total		830	1473	1016	1396	4715
Male	Time Categories	12 am-6:59 am	60	384	103	58	605
		7 am-8:59 am	143	178	177	246	744
		9 am-14:59 pm	434	370	370	565	1739
		3 pm-6:59 pm	312	285	247	247	1091
		7 pm-11:59 pm	148	436	208	140	932
	Total		1097	1653	1105	1256	5111
Total	Time Categories	12 am-6:59 am	94	545	159	105	903
		7 am-8:59 am	284	404	364	491	1543
		9 am-14:59 pm	736	784	762	1299	3581
		3 pm-6:59 pm	556	674	488	518	2236
		7 pm-11:59 pm	257	719	348	239	1563
	Total		1927	3126	2121	2652	9826

Figure 1. Methodology of the research

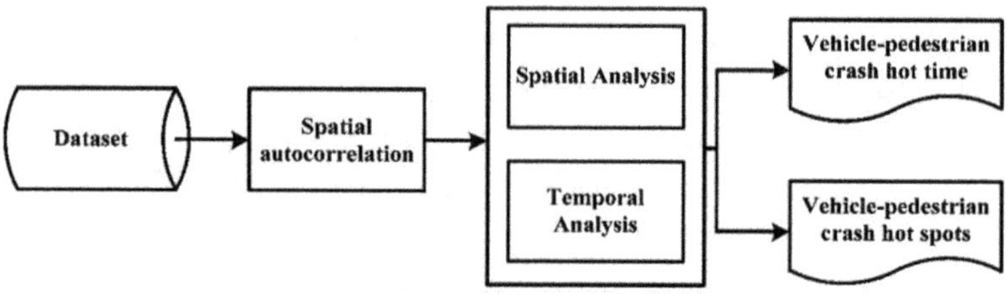

$$I = \frac{n}{S_0} \frac{\sum_{i=1}^{n} \sum_{j=1}^{n} w_{i,j} c_i c_j}{\sum_{i=1}^{n} c_i^2} \tag{1}$$

where, c_i is the deviation of feature i from its mean $(x_i - \overline{X})$, $w_{i,j}$ is the spatial weight between features i and j, n is equal to the number of features, and S_0 is the aggregate of all spatial weights:

$$S_0 = \sum_{i=1}^{n} \sum_{j=1}^{n} w_{i,j} \tag{2}$$

The z_i score that indicates whether or not we can reject the null hypothesis (there is no spatial clustering) is computed as:

$$z_i = \frac{I - E[I]}{\sqrt{V[I]}} \tag{3}$$

where:

$$\begin{aligned} E[I] &= {-1}/{(n-1)} \\ V[I] &= E[I^2] - E[I]^2 \end{aligned} \tag{4}$$

6.3.2. Identifying Areas With High Risk of Crash

KDE involves placing a symmetrical surface over each point and then evaluating the distance from the point to a reference location and then summing the value for all the surfaces for that reference location. This procedure is repeated for successive points. This allows us to place a kernel over each observation, and gives us the density estimate of the distribution of collision points. This surface has a maximum value at the reference point and this value decreases with increase in distance from the reference point and reaches zero at the radius distance from reference point (Pulugurtha et al, 2007). One common mathematical function used for KDE is:

$$f(x,y) = \frac{1}{nh^2} \sum_{i=1}^{n} K\left(\frac{d_i}{h}\right) \tag{5}$$

where $f\left(x,y\right)$ is the density estimation at location *(x,y)*, *n* is the number of observations, *h* is the bandwidth or kernel size or smooth parameter, *K* is the kernel function, and d_i is the distance between the location *(x,y)* and the location of ith observation.

Whereas in a simple density method, a circular neighbourhood is considered around each cell, in the kernel method, the research area is divided into predetermined number of cells. Thus, the kernel method draws a circular neighbourhood around each feature point (here each vehicle-pedestrian crash). There are different types of kernel functions, such as Gaussian, Quartic, Conic, negative exponential, and epanichnekov (Levine et al, 2002, Kuter et al, 2011). In this research, the Quartic kernel which is one of the three most common types of kernel functions, is applied (Schabenberger & Gotway, 2004, Xie & Yan, 2008). The specific form of the Quartic kernel function is:

$$K\left(\frac{d_i}{h}\right) = K\left(1 - \frac{d_i^2}{h^2}\right) \text{When } 0 < d_i \le h \tag{6a}$$

$$K\left(\frac{d_i}{h}\right) = 0 \text{ When } d_i > h \tag{6b}$$

In Equations 6a and 6b, *K* is the kernel function, and d_i is the distance between the location *(x,y)* and the location of ith observation. In Equation 6b, *K* is applied to ensure the total volume under Quartic curve is 1. The common values used for *K* include $\frac{3}{\pi}$ and $\frac{3}{4}$.

According to the literature, the accuracy of kernel function type in KDE, *K* is less important than the impact of the bandwidth *h* (Silverman, 1986, Schabenberger & Gotway, 2004, Xie & Yan, 2008, Loo et al, 2011, O'Sullivan & Unwin, 2014). Many studies have shown that the selection of the bandwidth or smoothing parameter in KDE is subjective (Bil et al, 2013). Different studies have selected different bandwidth values according to the area of study and size of the dataset. In general, according to Equation 6, selecting large bandwidth values $(h \to \infty)$ will decrease the density $(f(x,y) \to 0)$ and will show significant smoothing and low-density values (over smooth). In contrast, a small bandwidth value will result in less smoothing (under smooth), producing a map that depicts local variations in point densities (Chainey et al, 2002).

6.4. RESULTS

6.4.1. Spatial Autocorrelation

Spatial autocorrelation was applied to show the dependency between pedestrians' age, gender type, and the location of vehicle-pedestrian crashes. Table 2 shows the results of Global Moran's I for vehicle-pedestrian crashes for different pedestrian age groups and gender types. This table indicates that there are positive correlations between locations of vehicle-pedestrian crashes, and the age and gender of pedestrians.

These results for the z-score and Moran's I illustrate that the dependency between crash locations and pedestrians' age is not very significant for male pedestrians less than 18 years of age. This means that for male pedestrians in this age group, vehicle-pedestrian crashes are not significantly clustered. Apart from this group of pedestrians, the z-scores and p-values in the Moran's I analysis show that there is a less than 1% chance that any of the other clustered patterns may be the result of random chance. Furthermore, our results indicate that this dependency increases with age for male pedestrians for all age groups and increases for female pedestrians up to 65 years of age.

The results of spatial autocorrelation show that the age and gender of pedestrians have an influence on the location of vehicle-pedestrian crashes. Therefore, each age and gender type may have different crash hot spots that need to be explored.

6.4.2. Temporal Analysis

According to Table 1, male pedestrians are involved in slightly more vehicle-pedestrian crashes. Moreover, this table shows that most of the crashes occur between 9:00am and 2:59pm for both male and female pedestrians. For this period of time, pedestrians over 65 years of age are involved in more vehicle-pedestrian crashes, with 1,299 crashes. In addition, Table 1

Table 2. Global Moran's I spatial autocorrelation results

Age Groups	Male			Female		
	Moran's I	z-Score	p-Value	Moran's I	z-Score	p-Value
Under 18	0.006	1.439	0.15	0.026	5.916	<0.01
18-34	0.045	10.696	<0.01	0.058	13.512	<0.01
35-65	0.058	13.195	<0.01	0.066	15.669	<0.01
65+	0.074	16.274	<0.01	0.066	14.544	<0.01

in chapter 7 indicates that pedestrian between 18 and 34 and over 65 years have the highest total number of vehicle-pedestrian crashes, with 3,126 and 2,652 crashes respectively. Pedestrians between 18 and 34 years of age are more active and most likely to walk part of their journeys to or from work. This activity increases the risk of vehicle-pedestrian crashes for this age group. Furthermore, according to an Australian health survey in 2011-12, pedestrians over 65 years of age walk for fitness, recreation or sport more than other age groups (ABS 2014-2015). Moreover, older pedestrians usually walk more slowly than younger pedestrians and their levels of exposure to vehicular traffic may increase, which may increase their risk of experiencing a vehicle-pedestrian crash (Tarawneh, 2001).

Figure 2 shows the temporal distribution of vehicle-pedestrian crashes during weekdays and weekends, according to pedestrian gender types. This figure indicates that males and females have a similar temporal distribution of vehicle-pedestrian crashes during the weekdays. However, male vehicle-pedestrian crashes are more frequent than female crashes on weekends. Figure 2(b) shows that this difference between male and female vehicle-pedestrian crashes on weekends is more significant from 7:00pm to midnight.

6.4.2.1. Pedestrians Less Than 18 Years of Age

Figure 3(a) shows the temporal distribution of vehicle-pedestrian crashes for pedestrians less than 18 years of age. This age group includes school-aged pedestrians, and as this figure illustrates, there are two vehicle-pedestrian crash peaks around 8:00 am and 3:00 pm, when students are going to or leaving schools. Moreover, Figure 3(a) reveals that the pattern of vehicle-pedestrian crashes for males and females is very similar for this age group. Furthermore, this figure shows that the frequency of vehicle-pedestrian crashes is much

Figure 2. Temporal distribution of vehicle-pedestrian crashes for different gender types during weekdays and weekends

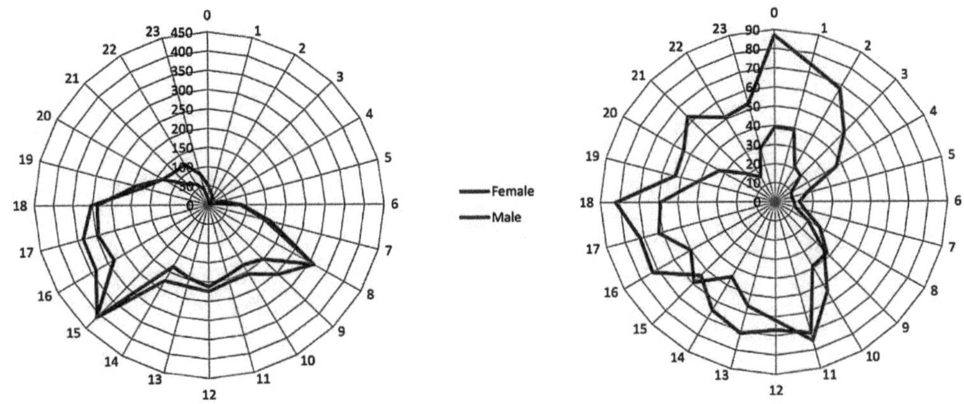

higher when they are leaving schools in the afternoon than when they are going to school in the morning.

6.4.2.2. Pedestrians Between 18 and 34 Years of Age

Figure 3(b) shows the distribution of vehicle-pedestrian crashes during the 24 hours of the day for male and female pedestrians between 18 and 34 years of age. This age group accounts for about 32% of vehicle-pedestrian crashes. According to this figure, male and female pedestrians in this age group have different patterns of crash times. For female pedestrians, crashes mainly occur at 8:00 am, and between 5:00 pm and 6:00 pm. However, for male pedestrians, in addition to vehicle-pedestrian crashes at 8:00 am and 6:00 pm, night-time crashes are also very significant. Figure 3(b) shows that the frequency of vehicle-pedestrian crashes for female pedestrians is higher than that for male pedestrians during the day, especially between 2:00 pm and 6:00 pm. However, after 8:00 pm, the frequency of male pedestrian crashes is significant and much higher than that for female pedestrians.

6.4.2.3. Pedestrians Between 35 and 64 Years of Age

Figure 3(c) shows the temporal distribution of vehicle-pedestrian crashes for pedestrians between 35 and 64 years of age. According to this figure, male and female crashes have very similar temporal distributions, especially after 2:00pm. Figure 3(c) also shows that there are four peaks for vehicle-

Figure 3. Temporal distribution of vehicle-pedestrian crashes for different pedestrian age groups and gender types

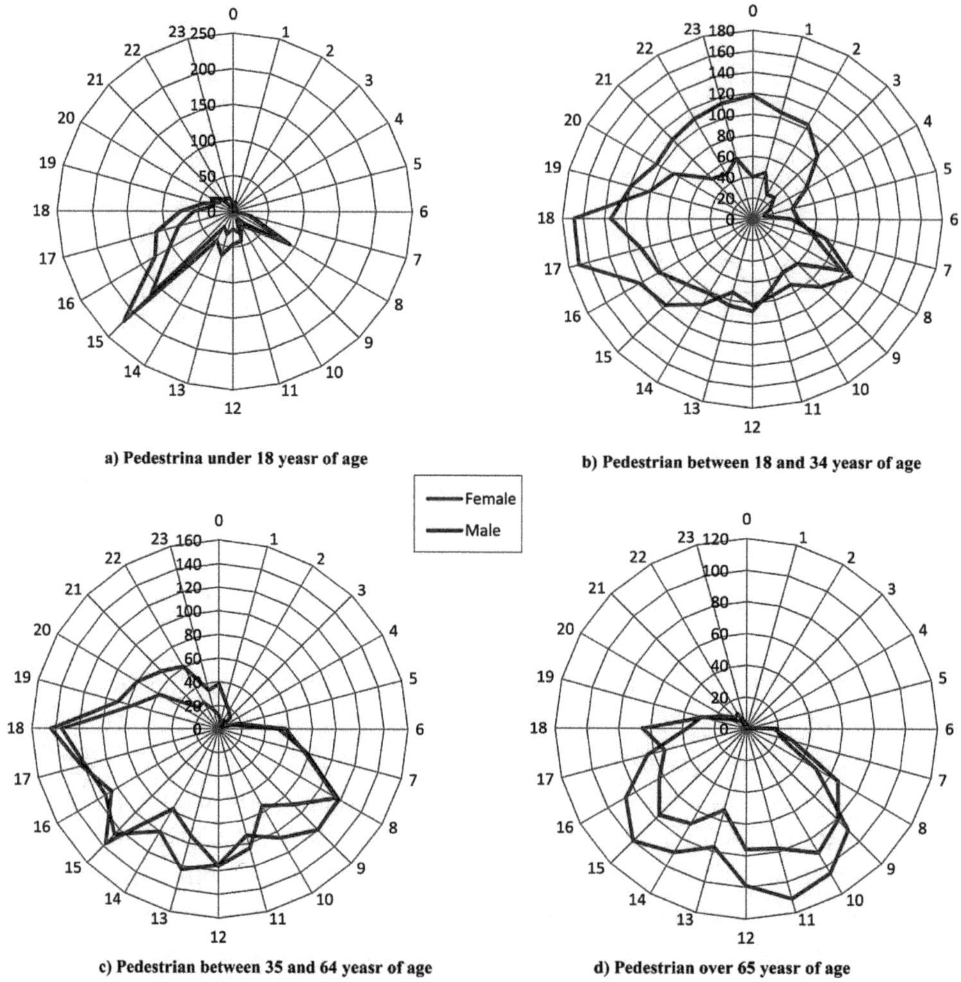

pedestrian crashes for this age group. However, the hot times are different for men and women. For men, 8:00am and 12:00am are the hot-times for vehicle-pedestrian crashes in the morning, whereas the hot time occurs between 8:00am and 9:00am for women. Moreover, 3:00pm and 6:00pm are other hot-times for male pedestrians in this age group, whereas the hot times in the afternoon occur at 1:00pm, 3:00pm, and 6:00pm for female pedestrians. Furthermore, Figure 3(c) shows that about 35% of pedestrian crashes occur between 10:00am and 3:00pm for this age group.

6.4.2.4. Pedestrians Over 65 Years of Age

About 16% of vehicle-pedestrian crashes are related to pedestrians over 65 years of age. According to Figure 3(d), about 50% of these crashes occur during the off-peak traffic period (between 10:00am and 3:00pm). Moreover, for this group of people, vehicle-pedestrian crashes occur more often on weekdays than on weekends (85% of crashes occur on weekdays).

6.4.2.5. Summary of Temporal Analysis

In summary, the results of temporal analysis show that different age groups and gender types have different temporal distributions for vehicle-pedestrian crashes. For pedestrians under 18 years of age, 8:00 am and 3:00 pm, when students are going to or leaving school, are the two hot times and the frequency of vehicle-pedestrian crashes for this age group is higher when they return home from school. The temporal distributions of vehicle-pedestrian crashes for pedestrians between 18 and 34 years of age show that male and female pedestrians have different hot times. For female pedestrians, crashes mainly occur at 8:00 am and between 5:00 pm and 6:00 pm. However, for male pedestrians, the frequency of crashes at night is also very significant. For pedestrians between 34 and 64 years of age, four crash peak times are identified for males and females. In the morning, 8:00 am and 9:00 am are crash hot times for male and female pedestrians, respectively. In addition, 12:00 pm for male pedestrians and 1:00 pm for female pedestrians, and 3:00 pm and 6:00 pm for both gender types are crash hot times. Finally, for pedestrians aged 65 and over, the frequency of vehicle-pedestrian crashes during the off-peak traffic period is significant for both gender types. Furthermore, crashes occur more often on weekdays than on weekends for this pedestrian age group.

6.4.3. Spatial Analysis

In this research, different bandwidth values were examined to find the most appropriate bandwidth value. The results of KDE analysis with different bandwidth values showed that KDE related to the 600m bandwidth value was the best. Therefore, 600m was selected as the KDE bandwidth value for hotspot analysis. The density of vehicle-pedestrian crashes is displayed by continuous surfaces in a raster map. In this map, which shows the results of

KDE, lighter shades represent locations with lower crash densities, while darker shades indicate areas with higher crash densities.

6.4.3.1. Spatial Distribution of Crashes During the Day and Night

The results of the spatial analysis reveal that vehicle-pedestrian crash hotspots are concentrated in the CBD and the streets around this area. Consistent with our temporal analyses, vehicle-pedestrian crashes occur more frequently at 8-9am and 3-6pm when people who work in the CBD commute to or from work (see Figure 4). However, Figures 4 and 5 reveal that vehicle-pedestrian crashes have different spatial patterns during the night and the day. Figures 4 and 5 show that there are several high-risk areas for vehicle-pedestrian crashes including the CBD during both day and night times, and also several other areas for the night-time. Moreover, Figures 6 and 7 show that these differences between the distribution of crashes during the day and night are similar for male and female vehicle-pedestrian crashes.

Figure 8 shows the distribution of bars and restaurants. In this figure, vehicle-pedestrian crashes during the night (7:00pm to 7:00am) are shown with black dots. Figure 8 indicates that most of the vehicle-pedestrian crashes occur in areas with a high density of restaurants, bars, clubs and liquor shops (dark shades in KDE results).

6.4.3.3. Hot Spots for Different Age Groups and Gender Types

Spatial analysis of vehicle-pedestrian crashes shows that the distributions of these crashes vary by the age group and gender of pedestrians. According to Figure 9, hotspots of vehicle-pedestrian crashes are more concentrated for female than male pedestrian crashes, especially for pedestrians under 18 and over 65 years of age. As shown in Table 2, the spatial correlation for vehicle-pedestrian crashes involving males who are less than 18 years of age is not significant. Therefore, the distribution of crashes for this age group and gender type is not very clustered and the results of the KDE show more areas with lower crash density (light grey area in KDE in Figure 9b). Figure 10 shows the distribution of schools and vehicle-pedestrian crashes for pedestrians less than 18 years of age. This figure reveals that the distribution of vehicle-pedestrian crashes for this age group is related to the distribution of schools.

6.4.3.3. Hot Spots for Different Age Groups and Gender Types

Spatial analysis of vehicle-pedestrian crashes shows that the distributions of these crashes vary by the age group and gender of pedestrians. According to Figure 9, hotspots of vehicle-pedestrian crashes are more concentrated for female than male pedestrian crashes, especially for pedestrians under 18 and over 65 years of age. As shown in Table 2, the spatial correlation for vehicle-pedestrian crashes involving males who are less than 18 years of age is not significant. Therefore, the distribution of crashes for this age group and gender type is not very clustered and the results of the KDE show more areas with lower crash density (light grey area in KDE in Figure 9b). Figure 10 shows the distribution of schools and vehicle-pedestrian crashes for pedestrians less than 18 years of age. This figure reveals that the distribution of vehicle-pedestrian crashes for this age group is related to the distribution of schools.

Moreover, the results of this research show that for adult pedestrians, the crash hotspots are distributed in different areas. For pedestrians between 18 and 65 years of age, most vehicle-pedestrian crashes occur in the CBD for both male and female pedestrians. This age group includes people of working age, and there are many offices, shops, and education centres in the CBD. Furthermore, the travel data from the Victorian Integrated Survey of Travel and Activity (VISTA, 2016) show that pedestrian traffic in the Melbourne CBD is higher than in other areas (see Figure 11). Figure 12 shows the pedestrian traffic volumes for different age groups in the Melbourne area. This figure indicates that pedestrian traffic for the age group between 18 and 65 years of age is greater than the other age groups. However, according to Figure 9(d)

Figure 4. Spatial distribution of vehicle-pedestrian crashes for male pedestrians and a) during the day b) during the night

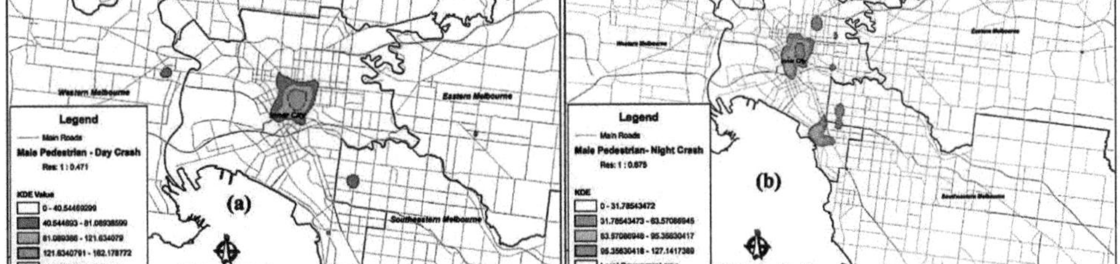

Figure 5. Spatial distribution of vehicle-pedestrian crashes for female pedestrians and a) during the day b) during the night

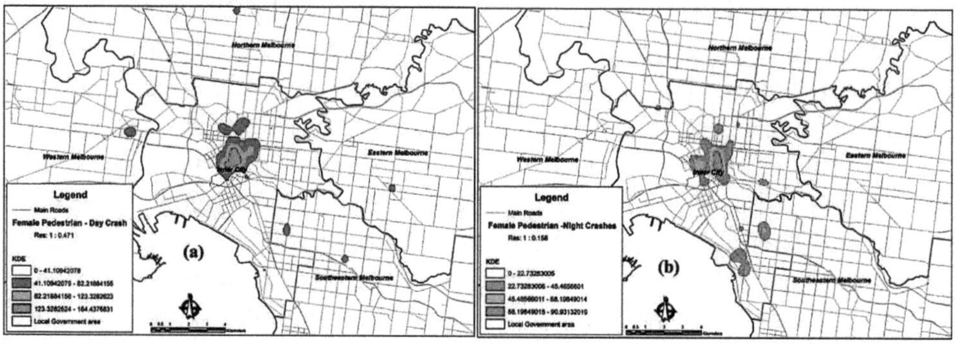

Figure 6. Distribution of bars, restaurants, and clubs and vehicle-pedestrian crashes during the night (7:00 pm-7:00 am)

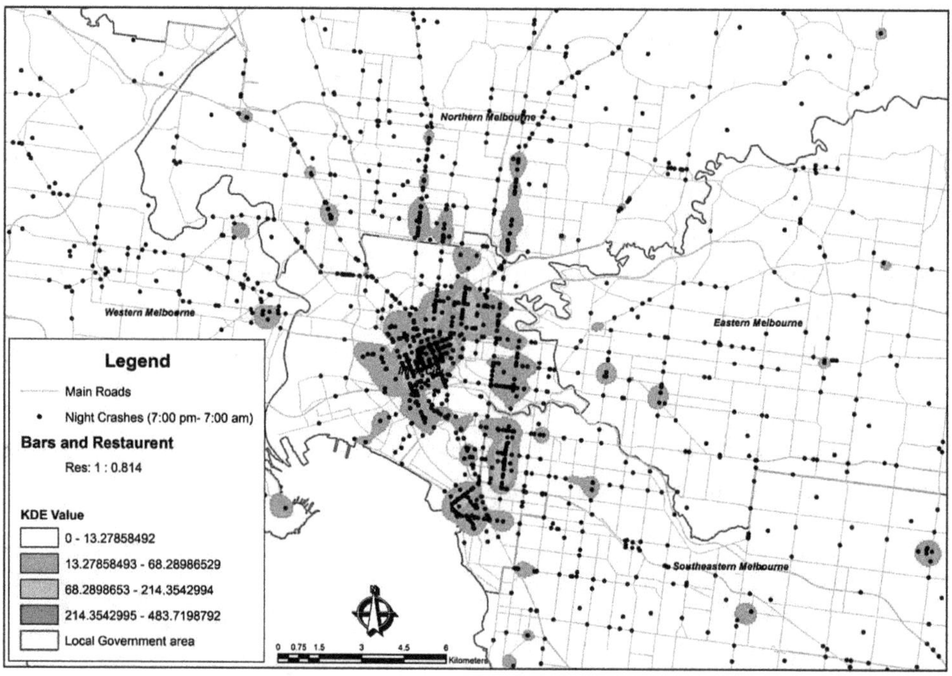

and 9(f), the distributions are different for male pedestrians. According to these two figures, there are some other crash hotspots for male pedestrians. A comparison of Figures 9(d) and (f) with Figure 8 shows that for male pedestrians between 18 and 65 years of age, areas around bars, restaurants and clubs are crash hotspots. This finding confirms that alcohol consumption may be an important contributing factor in vehicle-pedestrian crashes for men between 18 and 65 years of age. Furthermore, for pedestrians over 65 years of age, the Moran's I in Table 3 for male pedestrians shows that these crashes are significantly clustered, and in Figure 9(h), the KDE result shows more areas with high crash density for males than females.

6.4.3.4. Summary of Spatial Analysis

In summary, our spatial analyses show that age group and gender type influence the spatial distribution of vehicle-pedestrian crashes. For pedestrians between 16 and 65, vehicle-pedestrian crashes are concentrated around the CBD. However, these crashes are distributed among more areas for pedestrians under 18 and over 65 years of age. Moreover, the results of this research indicate that the distributions of vehicle-pedestrian crashes are different for male and female pedestrians during the night. For male pedestrians, areas around bars and restaurants are identified as hotspots, especially for the age group between 18 and 34 years of age. This finding shows that intoxication may be a significant contributing factor for this group of pedestrians. Schools and points of interest, such as community centres and parks, are likely hotspots for the under-18 and over-65 age groups, respectively.

6.5. DISCUSSION

The results of this research indicate that there are spatial dependencies between pedestrians' age and gender groups and the locations of crashes. This research shows that this dependency is greater for vehicle-pedestrian crashes at night. Therefore, different spider plots and KDE were then applied to explore the temporal and spatial distributions of vehicle-pedestrian crashes for different pedestrian age groups and gender types.

This research shows that 8:00am and 3:00pm are two hot times for pedestrians less than 18 years of age. Moreover, the results of this research show that the frequency of vehicle-pedestrian crashes is significantly higher

Figure 7. Spatial distribution of vehicle-pedestrian crashes for different pedestrian age groups and gender types

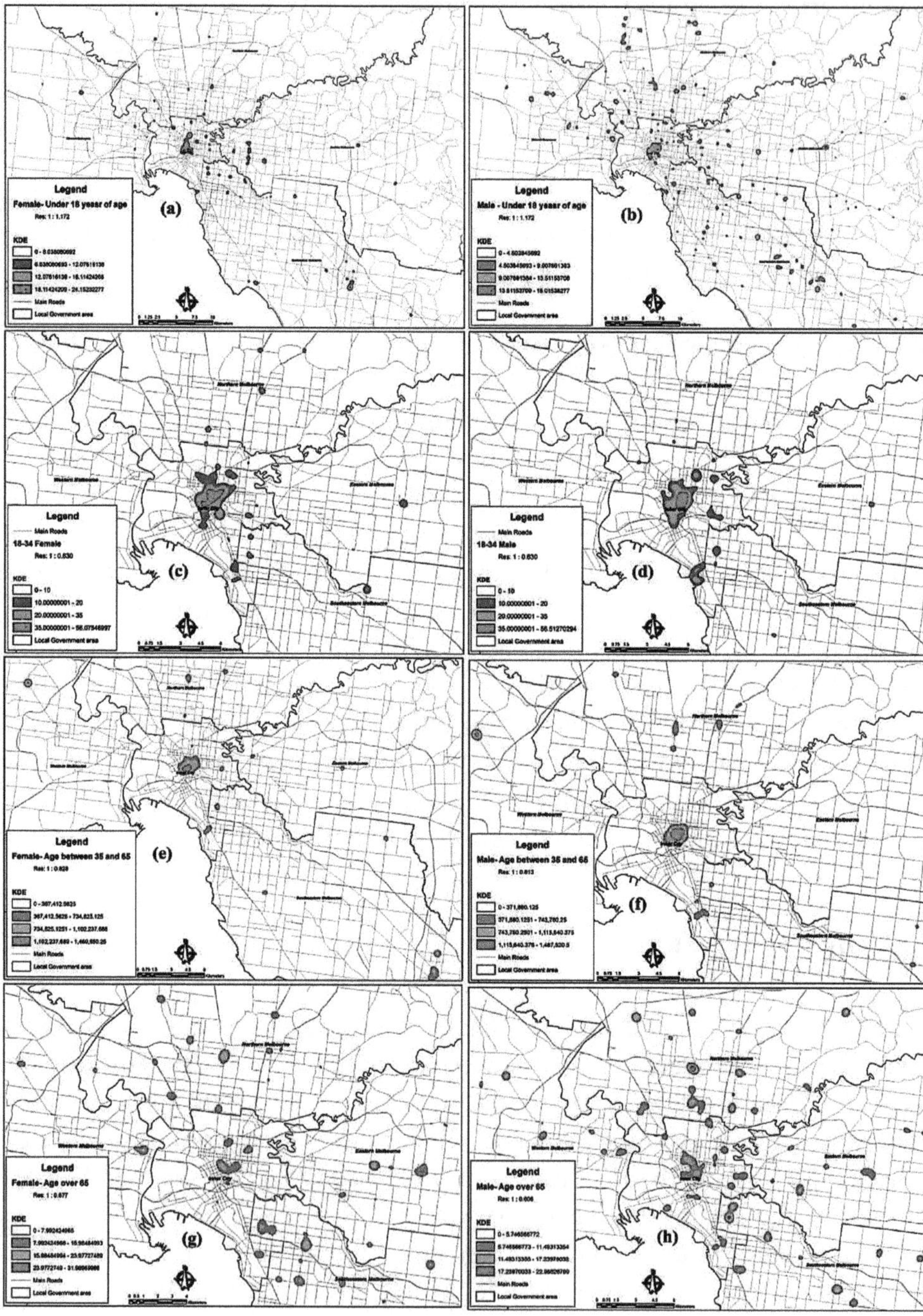

Figure 8. Spatial distribution of schools and vehicle-pedestrian crashes for pedestrians less than 18 years of age

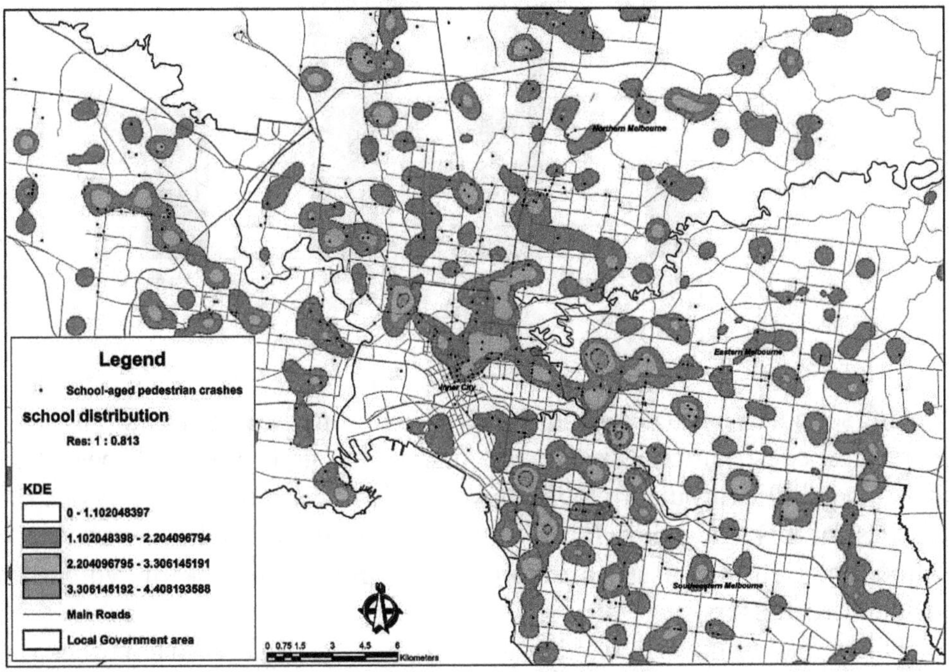

Figure 9. Walking travel volume to different Melbourne local government areas

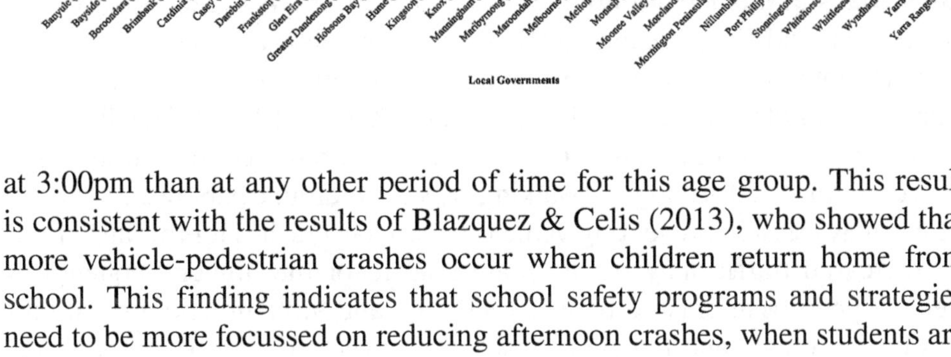

at 3:00pm than at any other period of time for this age group. This result is consistent with the results of Blazquez & Celis (2013), who showed that more vehicle-pedestrian crashes occur when children return home from school. This finding indicates that school safety programs and strategies need to be more focussed on reducing afternoon crashes, when students are

Figure 10. Walking travel volume for different age groups in Melbourne area.

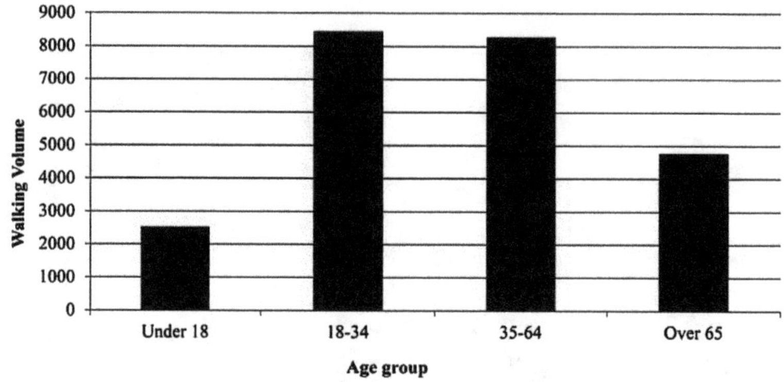

leaving schools. Controlling vehicle speed using engineering treatments, such as raised pedestrian crossings and roundabouts, and improving pedestrian crossing facilities around schools, may assist in preventing vehicle-pedestrian crashes for this age group. Furthermore, improving the safety awareness of school children may assist in improving the road safety of this vulnerable group of road users.

This research shows that although vehicle-pedestrian crash hot times vary for male and female pedestrians between 18 and 34 years of age, 8:00 am and 5:00 pm to 6:00 pm are common hot times for both genders. According to the Victorian traffic monitor reports for 2013, there are two traffic peak periods (7:30 am to 9:00 am and 4:30 pm to 6:00 pm) and one off-peak period (between 10:00 am and 3:00 pm) in the Melbourne metropolitan area (VicRoads, 2014). Therefore, pedestrian crashes at 8:00am and between 5:00pm and 6:00pm may be related to traffic peaks for pedestrians between 18 and 35 years of age. Moreover, the frequency of vehicle-pedestrian crashes is significant between 7:00 pm and 12:00 am for male pedestrians. Male pedestrians in this age group tend to stay out after work more than female pedestrians and therefore have an increased crash risk. Furthermore, the rate and amount of alcohol consumption for men in this group of age are greater than those of other groups (Wilsnack et al, 2009, AIHW, 2010), which may also increase their crash risk. The application of more control strategies to prevent intoxicated driving and walking, especially around bars, restaurants and clubs, may assist in preventing vehicle-pedestrian crashes for this age group.

Increased traffic may be the main reason for the higher frequency of vehicle-pedestrian crashes between 8:00 am and 9:00 am and around 6:00 pm for both male and female pedestrians between 35 and 64 years of age.

Nevertheless, the traffic off-peak period (10am – 3pm) is also a hazardous time for this pedestrian age group, and this result is consistent with the results of Plug et al. (2011), which show that the number of pedestrian crashes is still significant even when the traffic volume is not high. Some studies have shown that pedestrian volume and activity in the off-peak periods are higher than in peak periods, and this activity could increase the probability of vehicle-pedestrian crashes (Aultman-Hall et al, 2009, Miranda et al, 2011). Moreover, drivers tend to drive faster in off-peak periods. Therefore, speed may be a possible contributing factor in vehicle-pedestrian crashes during this time and increase the risk of this type of crash. Controlling the speed in this period of time and using different signs and traffic devices to warn drivers about pedestrians can assist in improving safety for this age group of pedestrians.

In addition, this research shows that weekdays and traffic off-peak periods are more important than other periods of time for pedestrians over 65 years of age. This result is consistent with the results of Nicaj et al. (2006), which show that most vehicle-pedestrian crashes involving pedestrians over 65 years of age occur between 10:00am and 5:00pm and on weekdays in New York. In addition, Figure 3(d) in chapter 7 reveals that there are more crashes involving female than male pedestrians in this age group. Bentley et al. (2010) found that female pedestrians spend more time walking around their local areas in the Melbourne metropolitan area. This difference between walking activity in men and women may change the risk of crashes and increase the risk for women. These findings may mean that older pedestrians, especially females, are more likely to be walking during the daytime on weekdays, especially between 10:00 am and 15:00 pm, than during the weekends or at other times of the day. Therefore, it is important to apply effective counter-measures, such as reducing speed limits or providing pedestrian crossing facilities, around POIs for older pedestrians, including health care centres, parks, shopping and social community centres, in order to improve the safety of these road users.

According to the results of this research, the CBD is identified as the main hotspot for vehicle-pedestrian crashes during the day. The existence of offices, shopping centres and educational centres in the CBD provides many origins and destinations for pedestrian trips. These POIs increase the number of trips and consequently the risks of vehicle-pedestrian crashes in these areas during the day. Furthermore, our results indicate that, in addition to the CBD, there are several other hotspots during the night for vehicle-pedestrian crashes. These hotspots are centred on restaurants, bars and clubs. This result is consistent with the results of other studies (Plug et al, 2011), which show different crash distributions between the day and night. Moreover, these results confirm the finding of Plug et al. (2011), which show that day-time crashes usually occur in the CBD and night-time crashes usually happen around cafés, restaurants, bars, shops, nightlife, and cultural events. This result is also similar to the results of DiMaggio et al. (2016) and Morrison et al. (2016), who found that an increase in the density of alcohol outlets in different suburbs increases the risk of crashes. Restricting the time for, and amount of, alcohol consumption, increasing pedestrian and driver education, installing warning signs and pedestrian barriers or fencing, and lowering speed limits at night in these hotspots may improve pedestrian safety in these areas and prevent vehicle-pedestrian crashes.

Moreover, this research finds different hotspots for different pedestrian age groups and genders. For pedestrians less than 18 years of age, areas around schools are risky areas. This finding is similar to the results of Abdel-Aty et al. (2007) and Blazquez and Celis (2013), which suggested that the majority of crashes involving school-aged children occur in areas near schools. In addition, the increased activities of pedestrians between 18 and 64 years of age in the CBD may have contributed to this area being identified as a vehicle-pedestrian crash hotspot during the day. Areas around restaurants, bars, and clubs are other vehicle-pedestrian crash hotspots for this age group during the night. Intoxication of pedestrians or drivers may be the main reason for these crashes. Moreover, this research identifies different crash hotspots for elderly pedestrians. These crash hotspots are clustered around POIs for this age group, such as parks and social community centres.

These results indicate that different safety strategies must be considered for different genders and age groups at different locations. For school-aged pedestrians, improving safety around schools (safe crossing guards, school zones, school areas, etc.) and increasing the traffic safety awareness of children when they are going to or leaving school may help to decrease the crash risk for this age group. For pedestrians over the age of 65, strategies targeting crash hotspots around parks, shops, and community centres during the day may be considered to decrease the risk of vehicle-pedestrian crashes for this age group. These measures include lowering the speed limit and installing more pedestrian crossings with longer crossing time.

For pedestrians between 18 and 65, safety measures can be concentrated in the CBD. These include anti-jaywalking education and enforcement campaigns, and installing median barriers, especially near tram stops. In addition, for pedestrians between 18 and 34 years of age, premises selling or serving alcohol, such as restaurants and bars, must be considered during the night. Strategies to manage alcohol consumption (e.g., restricting operating hours) and drink walking (e.g., installing roadside barriers in front of hotels), should be considered in these areas. Improving street lighting, providing warning signs for drivers, and lowering the speed limit may also reduce the risk of night-time vehicle-pedestrian crashes in these hotspots.

In general, reducing and controlling speed, especially around schools and POIs such as shopping centres or recreation areas, may assist in improving the safety of pedestrians. In addition, providing safe pedestrian crossing facilities around these locations can assist in reducing the number and severity of vehicle-pedestrian crashes. Moreover, applying more effective strategies to manage driver and pedestrian intoxication, and therefore driving or walking while intoxicated, is another approach to decrease the number and severity of vehicle-pedestrian crashes. Furthermore, education and awareness campaigns may be one way to engage the community and assist in reducing unsafe behaviour on the roads.

These results and recommendations are similar to current strategies that have been developed to improve vehicle-pedestrian safety in Victoria (2016). In this road safety strategy and action plan, reducing speed using different engineering treatments, such as pedestrian crossing treatments and raised platforms, is defined as an important strategy to reduce the number and severity of vehicle-pedestrian crashes. Likewise, strategies similar to those recommended in this study to reduce alcohol-related vehicle-pedestrian crashes are also included in this strategy and action plan.

6.6. CONCLUSION

In traffic safety, the location and time of vehicle-pedestrian crashes are known to be two important factors to consider when designing and applying safety strategies and counter-measures. In addition, the age and gender of pedestrians are important contributory factors for this type of crash. Pedestrians of different age groups and genders have diffident activity times and travelling behaviours. Therefore, the time and location of vehicle-pedestrian crashes can be different for different pedestrian ages and genders. This research examines the influence of pedestrian age group and gender type on the spatial and temporal distributions of vehicle-pedestrian crashes.

Spatial autocorrelation was applied in this research to identify the dependency between pedestrian age and gender groups and the location of vehicle-pedestrian crashes. These analyses showed that there is a significant dependency between age and gender groups and the location of vehicle-pedestrian crashes. The results of this research confirm that the spatial and temporal distributions of vehicle-pedestrian crashes differ for different pedestrian age and gender types. The results of this research indicate that vehicle-pedestrian crash hot times vary depending on pedestrian age group and gender type. Therefore, different safety policies and engineering strategies need to be applied for different age groups and gender types. For instance, active adult supervision for school-age pedestrians could assist them to navigate driveways, cars, roads and car parks safely. Likewise, improving pedestrian infrastructure facilities for older pedestrians, such as kerb extension and lowering of speed limits on streets and residential areas with large numbers of older pedestrians (community centres, clubs and health care centres), could improve their safety.

This research has revealed that vehicle-pedestrian crash hotspots differ during the day and night. The results of spatial analyses showed that the risks of vehicle-pedestrian crashes are significant around the CBD during the day. However, spatiotemporal analysis revealed that the existence of bars, clubs, and restaurants increases the probability of vehicle-pedestrian crashes during the night. Evidence-based road safety strategies, such as drink-driving and drink-walking enforcement, and improved street lighting, targeting these times and locations are required to improve pedestrian safety. Furthermore, this research shows that the influence of pedestrian age group and gender type on the spatial distributions of vehicle-pedestrian crashes and crash hotspots varies for different age groups and gender types.

REFERENCES

Abdel-Aty, M., Chundi, S. S., & Lee, C. (2007). Geo-spatial and log-linear analysis of pedestrian and bicyclist crashes involving school-aged children. *Journal of Safety Research*, *38*(5), 571–579. doi:10.1016/j.jsr.2007.04.006 PMID:18023642

ABS. (2011a). *Australian Bureau of Statistics*. Canberra, Australia: Australia, Year Book.

ABS. (2011b). *Australian Bureau of Statistics*. Available: www.abs.gov.au/

ABS. (2013). *Australia's population by country of birth Canberra*. Australian Bureau of Statistics.

ABS. (2015). *National Health Survey: First Results, 2014-15*. Canberra, Australia: Australian Bureau of Statistics 2015. Contract No.: No. 4364.0.55.001.

Aguero-Valverde, J., & Jovanis, P. P. (2006). Spatial analysis of fatal and injury crashes in Pennsylvania. *Accident; Analysis and Prevention, 38*(3), 618–625. doi:10.1016/j.aap.2005.12.006 PMID:16451795

AIHW. (2010). *Drinking patterns in Australia, 2001–2007*. Canberra: Australia: Australian Institute of Health and Welfare. Contract No.: Cat. no. PHE 133.

Al-Ghamdi, A. S. (2002). Pedestrian–vehicle crashes and analytical techniques for stratified contingency tables. *Accident; Analysis and Prevention, 34*(2), 205–214. doi:10.1016/S0001-4575(01)00015-X PMID:11829290

Al-Shammari, N., Bendak, S., & Al-Gadhi, S. (2009). In-Depth Analysis of Pedestrian Crashes in Riyadh. *Traffic Injury Prevention, 10*(6), 552–559. doi:10.1080/15389580903175313 PMID:19916125

Andrey, J., & Yagar, S. (1993). A temporal analysis of rain-related crash risk. *Accident; Analysis and Prevention, 25*(4), 465–472. doi:10.1016/0001-4575(93)90076-9 PMID:8357460

Aultman-Hall, L., Lane, D., & Lambert, R. (2009). Assessing Impact of Weather and Season on Pedestrian Traffic Volumes. *Transportation Research Record: Journal of the Transportation Research Board, 2140*(1), 35–43. doi:10.3141/2140-04

Bentley, R., Jolley, D., & Kavanagh, A. M. (2010). Local environments as determinants of walking in Melbourne, Australia. *Social Science & Medicine, 70*(11), 1806–1815. doi:10.1016/j.socscimed.2010.01.041 PMID:20299141

Bíl, M., Andrášik, R., & Janoška, Z. (2013). Identification of hazardous road locations of traffic accidents by means of kernel density estimation and cluster significance evaluation. *Accident; Analysis and Prevention, 55*(0), 265–273. doi:10.1016/j.aap.2013.03.003 PMID:23567216

Black, W. R. (1991). Highway accidents: A spatial and temporal analysis. *Transportation Research Record: Journal of the Transportation Research Board*, (1318), 75–82.

Blazquez, C. A., & Celis, M. S. (2013). A spatial and temporal analysis of child pedestrian crashes in Santiago, Chile. *Accident; Analysis and Prevention*, *50*, 304–311. doi:10.1016/j.aap.2012.05.001 PMID:22658462

Chainey, S., Reid, S., & Stuart, N. (2002). *When is a hotspot a hotspot? A procedure for creating statistically robust hotspot maps of crime*. London: Taylor & Francis.

DiMaggio, C., Mooney, S., Frangos, S., & Wall, S. (2016). Spatial analysis of the association of alcohol outlets and alcohol-related pedestrian/bicyclist injuries in New York City. *Injury Epidemiology*, *3*(1), 1. doi:10.118640621-016-0076-5 PMID:27747548

Elvik, R. (2009). The non-linearity of risk and the promotion of environmentally sustainable transport. *Accident; Analysis and Prevention*, *41*(4), 849–855. doi:10.1016/j.aap.2009.04.009 PMID:19540975

Fox, L., Serre, M. L., Lippmann, S. J., Rodríguez, D. A., Bangdiwala, S. I., Gutiérrez, M. I., ... Villaveces, A. (2015). Spatiotemporal Approaches to Analyzing Pedestrian Fatalities: The Case of Cali, Colombia. *Traffic Injury Prevention*, *16*(6), 571–577. doi:10.1080/15389588.2014.976336 PMID:25551356

Gómez, L. F., Parra, D. C., Buchner, D., Brownson, R. C., Sarmiento, O. L., Pinzón, J. D., ... Lobelo, F. (2010). Built Environment Attributes and Walking Patterns Among the Elderly Population in Bogotá. *American Journal of Preventive Medicine*, *38*(6), 592–599. doi:10.1016/j.amepre.2010.02.005 PMID:20494235

Henary, B. Y., Ivarsson, J., & Crandall, J. R. (2006). The Influence of Age on the Morbidity and Mortality of Pedestrian Victims. *Traffic Injury Prevention*, *7*(2), 182–190. doi:10.1080/15389580500516414 PMID:16854713

Holland, C., & Hill, R. (2007). The effect of age, gender and driver status on pedestrians' intentions to cross the road in risky situations. *Accident; Analysis and Prevention*, *39*(2), 224–237. doi:10.1016/j.aap.2006.07.003 PMID:16979132

Hosseinpour, M., Prasetijo, J., Yahaya, A. S., & Ghadiri, S. M. R. (2013). A Comparative Study of Count Models: Application to Pedestrian-Vehicle Crashes Along Malaysia Federal Roads. *Traffic Injury Prevention*, *14*(6), 630–638. doi:10.1080/15389588.2012.736649 PMID:23859313

Levine, N., Kim, K. E., & Nitz, L. H. (1995). Spatial analysis of Honolulu motor vehicle crashes: I. Spatial patterns. *Accident; Analysis and Prevention*, *27*(5), 663–674. doi:10.1016/0001-4575(95)00017-T PMID:8579697

Li, L., Zhu, L., & Sui, D. Z. (2007). A GIS-based Bayesian approach for analyzing spatial–temporal patterns of intra-city motor vehicle crashes. *Journal of Transport Geography*, *15*(4), 274–285. doi:10.1016/j.jtrangeo.2006.08.005

Loo, B. P. Y., & Tsui, M. K. (2005). Temporal and spatial patterns of vehicle-pedestrian crashes in busy commercial and shopping areas: A case study of hong kong. *Asian Geographer*, *24*(1-2), 113–128. doi:10.1080/10225706.2005.9684124

Miranda-Moreno, L. F., Morency, P., & El-Geneidy, A. M. (2011). The link between built environment, pedestrian activity and pedestrian–vehicle collision occurrence at signalized intersections. *Accident; Analysis and Prevention*, *43*(5), 1624–1634. doi:10.1016/j.aap.2011.02.005 PMID:21658488

Miranda, S., Carrasco, Y., & Jorge, G. (2011). *Pedestrian Volume Studies: A case study in the city of Gothenburg*. Göteborg, Sweden: Chalmers University of Technology.

Morrison, C., Ponicki, W. R., Gruenewald, P. J., Wiebe, D. J., & Smith, K. (2016). Spatial relationships between alcohol-related road crashes and retail alcohol availability. *Drug and Alcohol Dependence*, *162*, 241–244. doi:10.1016/j.drugalcdep.2016.02.033 PMID:26968094

NHTSA, . (2015). Traffic Safety Facts, 2012 Data: Pedestrians. *Annals of Emergency Medicine*, *65*(4), 452. doi:10.1016/j.annemergmed.2015.02.019

Nicaj, L., Wilt, S., & Henning, K. (2006). Motor vehicle crash pedestrian deaths in New York City: The plight of the older pedestrian. *Injury Prevention*, *12*(6), 414–416. doi:10.1136/ip.2005.010082 PMID:17170193

O'Sullivan, D., & Unwin, D. J. (2014). *Geographic information analysis*. Hoboken, NJ: John Wiley & Sons.

Pink, B. (2010). *Year Book Australia*. Contract No.: ABS Catalogue No. 1301.0.

Plug, C., Xia, J. C., & Caulfield, C. (2011). Spatial and temporal visualisation techniques for crash analysis. *Accident; Analysis and Prevention*, *43*(6), 1937–1946. doi:10.1016/j.aap.2011.05.007 PMID:21819821

Pulugurtha, S. S., Krishnakumar, V. K., & Nambisan, S. S. (2007). New methods to identify and rank high pedestrian crash zones: An illustration. *Accident; Analysis and Prevention*, *39*(4), 800–811. doi:10.1016/j.aap.2006.12.001 PMID:17227666

Schneider, R. J., Ryznar, R. M., & Khattak, A. J. (2004). An accident waiting to happen: A spatial approach to proactive pedestrian planning. *Accident; Analysis and Prevention*, *36*(2), 193–211. doi:10.1016/S0001-4575(02)00149-5 PMID:14642874

Siddiqui, C., Abdel-Aty, M., & Choi, K. (2012). Macroscopic spatial analysis of pedestrian and bicycle crashes. *Accident; Analysis and Prevention*, *45*, 382–391. doi:10.1016/j.aap.2011.08.003 PMID:22269522

Silverman, B. W. (1986). *Density estimation for statistics and data analysis*. London: CRC press. doi:10.1007/978-1-4899-3324-9

Sundquist, K., Eriksson, U., Kawakami, N., Skog, L., Ohlsson, H., & Arvidsson, D. (2011). Neighborhood walkability, physical activity, and walking behavior: The Swedish Neighborhood and Physical Activity (SNAP) study. *Social Science & Medicine*, *72*(8), 1266–1273. doi:10.1016/j.socscimed.2011.03.004 PMID:21470735

Tarawneh, M. S. (2001). Evaluation of pedestrian speed in Jordan with investigation of some contributing factors. *Journal of Safety Research*, *32*(2), 229–236. doi:10.1016/S0022-4375(01)00046-9

Tay, R., Choi, J., Kattan, L., & Khan, A. (2011). *A multinomial logit model of pedestrian–vehicle crash severity*. Academic Press.

Toran Pour, A., Moridpour, S., Tay, R., & Rajabifard, A. (2017). Spatial and Temporal Distribution of Pedestrian Crashes in Melbourne Metropolitan Area. *Road and Transport Research Journal*, *26*(1), 6-22.

Van Dyck, D., Cardon, G., Deforche, B., Sallis, J. F., Owen, N., & De Bourdeaudhuij, I. (2010). Neighborhood SES and walkability are related to physical activity behavior in Belgian adults. *Preventive Medicine*, *50*(Supplement), S74–S9. doi:10.1016/j.ypmed.2009.07.027 PMID:19751757

VICROADS. (2015). *Principal Bicycle Network (PBN)*. Victoria, Australia: Victorian Government Data Directory.

VicRoads. (2014). *Traffic monitor 2012-13*. Melbourne: Roads Corporation of Victoria.

WHO. (2013). *Pedestrian safety: a road safety manual for decision-makers and practitioners*. World Health Organization.

Wilsnack, R. W., Wilsnack, S. C., Kristjanson, A. F., Vogeltanz-Holm, N. D., & Gmel, G. (2009). Gender and alcohol consumption: Patterns from the multinational GENACIS project. *Addiction (Abingdon, England)*, *104*(9), 1487–1500. doi:10.1111/j.1360-0443.2009.02696.x PMID:19686518

Xie, Z., & Yan, J. (2008). Kernel Density Estimation of traffic accidents in a network space. *Computers, Environment and Urban Systems*, *32*(5), 396–406. doi:10.1016/j.compenvurbsys.2008.05.001

Chapter 7
Contributing Factors on Vehicle-Pedestrian Crash Severity of School-Aged Pedestrians

ABSTRACT

Every year, about 19% of vehicle-pedestrian crashes in Melbourne metropolitan area, Australia, involve pedestrians less than 18 years of age or school-aged pedestrians. This chapter aims to identify contributing factors on vehicle-pedestrian crash severity of this age group. Reasonable walking distance to schools is applied in geographic information systems (GIS) to identify vehicle-pedestrian crashes around schools. Then boosted decision tree (BDT) and cross-validation (CV) technique are applied to explore significant factors. Results show that the distance of pedestrians from school is a significant factor on vehicle-pedestrian crash severity for this age group. This result could assist in identifying a safe distance and safe zone around schools. Furthermore, public health indicators such as income and commuting type from or to school are found as other contributing factors to this crash type.

DOI: 10.4018/978-1-5225-7943-4.ch007

7.1. INTRODUCTION

To reduce the involvement of school children in such crashes, we need to identify contributing factors on vehicle-pedestrian crashes for this age group and improve safety in the school zones. In recent years, many research conducted to identify contributing factors on vehicle-pedestrian crash frequency and severity (Li et al, 2016, Toran Pour et al, 2016, Rifaat et al, 2017). For instance, Tay et al. (2011) identified that pedestrians' and drivers' age, and driving speed could influence vehicle-pedestrian crash severity in South Korea. Furthermore, in another study Toran Pour et al. (2017) indicated that neighbourhood social characteristics were as important as traffic and infrastructure variables in severity of pedestrian crashes. However, there is relatively fewer research focused on school-aged vehicle-pedestrian crashes. For instance, In addition, Graham and Glaister (2003) and Graham et al. (2005) found that the probability of child pedestrian casualties is higher in more deprived areas. In another study, Noland and Quddus (2004) found that more severe pedestrian injuries are associated with the areas with lower income, higher percent of local roads, higher per capita expenditure on alcohol, and larger numbers of people. Abdel-Aty et al. developed a GIS base crash analysis and Log-linear model for pedestrians under 19 years of age in Florida (Abdel-Aty et al, 2007). In this research, they showed that majority of school-aged children crashes occurred in the areas near schools. Furthermore, in this study drivers' and pedestrians' age, road geometry, speed limit, and speed ratio were also found to be correlated with the frequency of crashes. Also, in another study it is identified that child pedestrian crashes are more strongly associated with built environment features (Rothman et al, 2014).

Koopmans et al. (2015) investigated the vehicle-pedestrian injury crashes for pedestrians under 19 years of age in Chicago and found that environmental conditions such as weather condition, light, and location of crashes are contributing factors on crash injury severity of pedestrians. Lee et al. (2016) applied standard negative binomial and zero-inflated negative binomial models to identify the influencing environmental attributes of intersections on crashes involving children aged 10 to12 years of age in Korea near elementary schools. They found that a higher number of student crossings, a wider road width, the presence of crosswalks, student-friendly facilities at the intersection, and four-way intersections were significant and positively associated with perceived crash risk among school-aged children. Literature review shows that there are relatively few studies with an emphasis on more refined spatial distribution of school-aged crashes, particularly in the areas surrounding schools. Furthermore, the existing research applied linear buffer around schools to identify crashes and in the existing studies walking distance is not considered. The objectives of this study are:

1. To identify contributing factors on vehicle-pedestrian crash severity for the school-aged pedestrians. For this reason, the Boosted Decision Tree (BDT) is developed and influencing factors on this crash types identified.
2. To identify crash severity risk distance from schools using GIS analysis and BDT model development. Using Network analysis in GIS and Partial Dependence Plots in BDT model are developed to identify the in which distance from schools the severity of crashes in lower.

This paper is structured as follows. The next section of the paper presents the dataset and methodology of this research. The results are presented and discussed in Section 7.3. Finally, the outcomes are summarised in Section 7.4.

7.2. DATA AND METHODS

7.2.1. Dataset

To investigate the variables contributing school aged vehicle-pedestrian crash severity, data for these crashes on public roadways of Melbourne metropolitan area from 2004 to 2013 are extracted from RCIS. Of the total of 11,548 vehicle-pedestrian crashes, 2,161 are related to pedestrians less than 19 years of age. According to VicRoads severity classification, of the 2,161 school-aged vehicle-pedestrian crashes included in the study, 1.2% were fatal crashes, 43.5% were serious injury crashes, and. 55.3% were minor injury crashes. In addition to the crash data, data on the neighbourhood social and economic characteristics are extracted from the Australian Bureau of Statistics (2013). ArcMap GIS 10.3 is used to extract the social and economic variables related to each suburb where the corresponding vehicle-pedestrian collision occurred. ArcMap GIS 10.3 is also used to extract the traffic volume data from the Melbourne road network database for each crash location. In addition, to identify the distance of crashes from schools, 1274 schools including primary, secondary, language and special schools which are in Melbourne metropolitan area are used in GIS. Tables 1 and 2 show a summary of the categorical and continuous variables used in this study, respectively.

7.2.2. Methods

In this research network analysis in ArcMap 10.3 is used to identify distance between crash locations and schools. Then this distance with other factors are applied in Boosted Decision Tree Model (BDT) to identify contributing factors on school-aged vehicle-pedestrian crash severity. Furthermore, Partial Dependence Plot (PDP) is applied to find crash severity risk distance from schools and this result is used in GIS to identify roads with higher probability of crash severity around schools (Please see Chapter 4 for more detail about BDT and PDP).

7.3. RESULTS AND DISCUSSION

Figure 1 shows the top 10 most important predictor variables for school-aged vehicle-pedestrian crash severity. As shown in this figure, 'Crash Distance to School' is the most important contributing factor on the severity of school-aged vehicle-pedestrian crashes, showing that this variable is a significant influencing variable on vehicle-pedestrian crash severity at this age group. The results from this study show that this factor needs to be considered in vehicle-pedestrian crash studies around schools and for school-aged pedestrians.

Table 1. Categorical explanatory variables applied in BDT model

Variables		Percent
Severity	Fatal Crash	0.6
	Serious Injury Crash	37.5
	Minor Injury Crash	61.9
Day of Crash (school days)	Monday	18.3
	Tuesday	23.1
	Wednesday	18.4
	Thursday	19.7
	Friday	20.5
Time of Crash	Firs Peak (7:00-9:00 am)	32.7
	Off-Peak (10:00 am – 3:00 pm)	67.3

continued on following page

Table 1. Continued

Variables		Percent
Month of Crash	January	4.8
	February	8.8
	March	10.4
	April	7.9
	May	11.4
	June	7.7
	July	7.9
	August	8.6
	September	9.8
	October	8.2
	November	7.2
	December	7.4
Light Condition	Day	98.1
	Dusk/Dawn	1.5
	Other	0.4
Node Type	Intersection	49.5
	Mid-Blocks	49.8
	Other	0.7
Surface Condition	Dry	87.6
	Wet	9.1
	Other	3.3
Atmosphere Condition	Clear	90.4
	Rainy	5.8
	Other	3.8
Pedestrian Gender	Male	54.2
	Female	44.7
Driver Gender	Male	47.5
	Female	43.8
Speed Limit	Under 50 km/h	49.6
	60-70 km/h	41.8
	>70 km/h	5.5
	Other	3.1

Median income of crash location neighbourhoods is the second contributing factor on school-aged vehicle-pedestrian crash severity. This result is similar to the other research that found income could influence the vehicle-pedestrian crashes around schools (LaScala et al, 2000, LaScala et al, 2004). This result will assist transportation engineers, planners and policy makers in identifying the target customer segments for improving child pedestrian safety. Knowing the right target audience is critical for the success of safety education and communications programs, such as publishing child pedestrian safety bulletins, safety programs at schools, and using warning messages on billboards.

According to Figure 1, Traffic volume and Distance of crash to public transport stops are two next contributing factors on the vehicle-pedestrian crash severity of this age group. These results are consistent with results from other studies that found traffic volume and public transport stops could influence vehicle-pedestrian crashes (Assailly, 1997). Furthermore, Figure 1 indicates that type of commuting to/from school could be a contributing factor on school-aged vehicle-pedestrian crash severity. School-home commuting type could influence walking distance and exposure of school-aged pedestrians to vehicular traffic. Finally, this figure shows that month of crash (March) is another factor that is identified as influencing factor on this type of crashes. In Australia, March is the last month of school term 1 and it is one of the busiest months for schools and school-aged pedestrians.

Figure 2 shows the partial dependence plots for the first 10 top factors for different levels of vehicle-pedestrian crash severity. In this figure, it is possible to identify the influence of different variables on vehicle-pedestrian crash severity levels. Figure 2(a) illustrates that increasing the distance of crash location from/to schools from about 450 meters will increase the probability of fatal crashes. Therefore, it is possible to define this distance as safe walking distance toward schools for school-aged pedestrians. Furthermore, this result assists in identifying target roads to apply road safety strategies and plans for

Table 2. Descriptive statistics for continuous variables applied in BDT model

Variable	Unit	Mean	Std. Deviation
Traffic Volume	Vehicle per day	11694.3	8973.8
Distance to school	Meters	595.2	466.3
Distance to Public Transport Stops	Meters	133.1	169.0
Median income	AUS Dollars	599	155.6
Use Public Transport	Percent	10.4	6.5
Use other type of transport	Percent	3.3	2.3
Use Private	Percent	62.5	14.6
Use Walk	Percent	4.1	7.8
Use Multimodal	Percent	5.1	1.5

Figure 1. Top 10 relative importance of predictor variables for school-aged vehicle-pedestrian crashes in BDT model

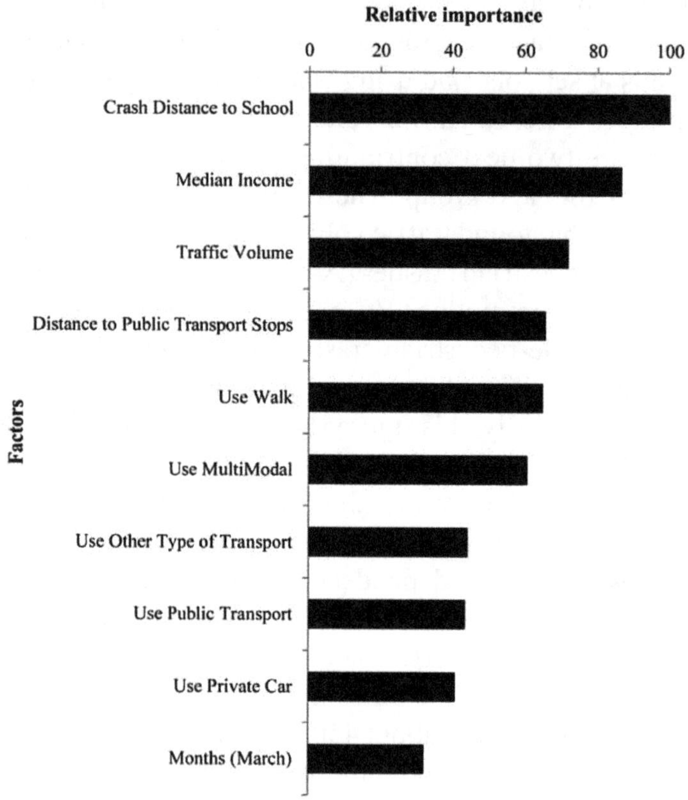

school-aged pedestrians including speed calming, warning systems or child pedestrian crossing supervision.

Figure 2(b) presents the influence of median income on crash severity of school-aged pedestrians around the schools. This figure shows that the probability of a vehicle-pedestrian crash to be fatal is much higher in suburbs around schools with less weekly income. According to this figure, the risk of fatal crashes that occurs in suburbs with median weekly income between $370.00 and $450.00 per week (low income) could be more than other suburbs. This result is consistence with the results from other studies that showed the probability of vehicle-pedestrian crashes to be fatal is more in low-income suburbs (Zhu & Lee, 2008). Our results, as highlighted in Figure 2(b), suggest that suburbs with low median income could be targeted for school-aged pedestrian safety educational programs or campaigns. These programs could increase the traffic safety knowledge, especially safe walking knowledge, and improve pedestrian safety for child living in these targeted suburbs.

Moreover, Figure 2(c) shows that increase in traffic volume from about 15,000 to 19,000 vehicles per day in roads around schools could increase the risk of fatal crash. The risk of fatal crash decreased and then remained stable after 21,000 vehicles per day. The results of the present research are consistent with the results of other studies showing that increasing traffic volume can increase pedestrian crash frequency and the probability of pedestrian crash severity (Morency, 2012, Toran Pour et al, 2017). These results suggest that transportation engineers and planners may want to target roads with more than 15,000 vehicles per day to improve the safety of these vulnerable road users around schools. More pedestrian crossings, pedestrian signals and flashing lights on these roads may assist in improving the safety of school-aged pedestrians.

Figures 2(d) shows that the severity of vehicle-pedestrian crashes for this age group and around schools is decreased by increase in the distance of crash to/from public transport stops (from 0 to 50 meters) and then remained stable (from 50 to 110 meters). According to this figure the probability of fatal crashes then increases with the increase in the distance of pedestrian crash locations to public transport stops from 160 to 450m. Using different warning signals and signs around public transport stops may increase the attention given by drivers to pedestrians. More research is required to analyse vehicle-pedestrian crashes in the vicinity of public transport stops and further the 450m distance to identify appropriate pedestrian safety programs (Toran Pour et al, 2017).

Figures 2(e) to (i) show the influence of using different transport type to commute between home and schools on school-aged vehicle-pedestrian crash severity. These figures show that different commute type could influence the vehicle-pedestrian crashes differently. For instance, increasing the percentage of people that walk or use public transport to commute between home and schools could slightly increase the risk of fatal vehicle-pedestrian crashes. However, using multimodal transport (e.g. Private-public transport) could decrease this risk.

Finally Figure 2(j) shows that the risk of school-aged vehicle-pedestrian crashes could vary at different months of year. This figure shows that the risk of fatal crash is increased in March. In Victoria, Australia, March is the last mount of term 1 for schools and in this month, students may be more active than other months. This result could assist in identifying the month

Figure 2. Partial dependence plot for first 10 top contribution factors to the school-aged vehicle-pedestrian crash severity

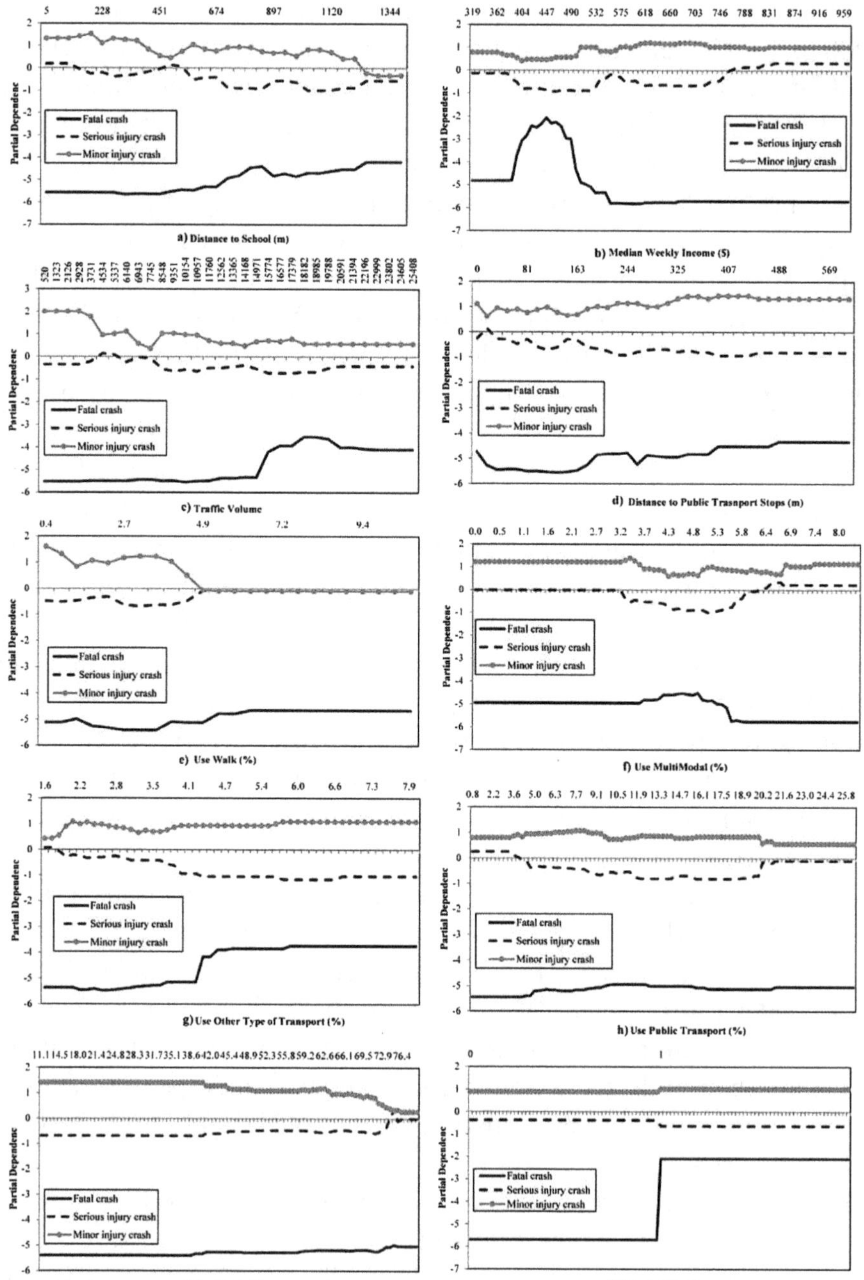

with higher risk of accident to apply safety programs for students and parents to improve the safety of school-aged pedestrians.

7.4. CONCLUSION

Identifying contributing factors on school-aged vehicle-pedestrian crash severity could assist transportation engineers, road safety professionals and policy makers in developing and implementing effective countermeasures around schools to reduce the number of pedestrian deaths and injuries of these vulnerable road users. In this research, the BDT model was applied to identify the contributing factors on the school-aged vehicle-pedestrian crashes. Also, GIS is applied to identify the distance of crash location to/from schools and extract socio-economic factors such as income and commute type related to location of crashes. Results from this research would provide valuable information to assist road safety professional in targeting the right neighbourhoods to implement different safety measures related to pedestrians and drivers, as well as targeting site specific safety measures to reduce vehicle-pedestrian crashes for school-aged pedestrians.

This study found that distance of crash to schools is the most important variable in vehicle-pedestrian crash severity around schools. Moreover, this research revealed that public wellbeing indicators such as median income and using public transport have influence on severity of crashes in this age group. This research found that traffic volume and distance of crashes to/from public transport stop are two other contributing factors on school-aged vehicle-pedestrian crashes in Melbourne metropolitan area. Furthermore, this research showed that month of the year could be an important factor for vehicle-pedestrian crash severity in this age group. These results could assist transport and safety planners in identifying target suburbs/roads and introducing appropriate countermeasures such as pedestrian safety educational programs to improve the safety of school-aged pedestrians.

REFERENCES

Abdel-Aty, M., Chundi, S. S., & Lee, C. (2007). Geo-spatial and log-linear analysis of pedestrian and bicyclist crashes involving school-aged children. *Journal of Safety Research*, *38*(5), 571–579. doi:10.1016/j.jsr.2007.04.006 PMID:18023642

ABS. (2011a). *Australian Bureau of Statistics*. Canberra, Australia: Australia, Year Book.

ABS. (2011b). *Australian Bureau of Statistics*. Available: www.abs.gov.au/

ABS. (2013). *Australia's population by country of birth*. Australian Bureau of Statistics.

ABS. (2015). *National Health Survey: First Results, 2014-15*. Canberra, Australia: Australian Bureau of Statistics 2015. Contract No.: No. 4364.0.55.001.

Assailly, J.-P. (1997). Characterization and prevention of child pedestrian accidents: An overview. *Journal of Applied Developmental Psychology*, *18*(2), 257–262. doi:10.1016/S0193-3973(97)90039-3

Graham, D., Glaister, S., & Anderson, R. (2005). The effects of area deprivation on the incidence of child and adult pedestrian casualties in England. *Accident; Analysis and Prevention*, *37*(1), 125–135. doi:10.1016/j.aap.2004.07.002 PMID:15607283

Graham, D. J., & Glaister, S. (2003). Spatial variation in road pedestrian casualties: The role of urban scale, density and land-use mix. *Urban Studies (Edinburgh, Scotland)*, *40*(8), 1591–1607. doi:10.1080/0042098032000094441

Koopmans, J. M., Friedman, L., Kwon, S., & Sheehan, K. (2015). Urban crash-related child pedestrian injury incidence and characteristics associated with injury severity. *Accident; Analysis and Prevention*, *77*, 127–136. doi:10.1016/j.aap.2015.02.005 PMID:25703351

LaScala, E. A., Gerber, D., & Gruenewald, P. J. (2000). Demographic and environmental correlates of pedestrian injury collisions: A spatial analysis. *Accident; Analysis and Prevention*, *32*(5), 651–658. doi:10.1016/S0001-4575(99)00100-1 PMID:10908137

LaScala, E. A., Gruenewald, P. J., & Johnson, F. W. (2004). An ecological study of the locations of schools and child pedestrian injury collisions. *Accident; Analysis and Prevention*, *36*(4), 569–576. doi:10.1016/S0001-4575(03)00063-0 PMID:15094409

Lee, G., Park, Y., Kim, J., & Cho, G.-H. (2016). Association between intersection characteristics and perceived crash risk among school-aged children. *Accident; Analysis and Prevention*, *97*, 111–121. doi:10.1016/j.aap.2016.09.001 PMID:27612169

Li, D., Ranjitkar, P., Zhao, Y., Yi, H., & Rashidi, S. (2016). Analyzing pedestrian crash injury severity under different weather conditions. *Traffic Injury Prevention.*

Morency, P., Gauvin, L., Plante, C., Fournier, M., & Morency, C. (2012). Neighborhood Social Inequalities in Road Traffic Injuries: The Influence of Traffic Volume and Road Design. *American Journal of Public Health, 102*(6), 1112–1119. doi:10.2105/AJPH.2011.300528 PMID:22515869

Noland, R., & Quddus, M. (2004). Analysis of pedestrian and bicycle casualties with regional panel data. *Transportation Research Record: Journal of the Transportation Research Board*, 28–33.

Rifaat, S.M., Tay, R., Raihan, S.M., Fahim, A., & Touhidduzzaman, S.M. (2017). Vehicle-Pedestrian crashes at Intersections in Dhaka city. *The Open Transportation Journal, 11*(1).

Rothman, L., Macarthur, C., To, T., Buliung, R., & Howard, A. (2014). Motor Vehicle-Pedestrian Collisions and Walking to School: The Role of the Built Environment. *Pediatrics, 133*(5), 776–784. doi:10.1542/peds.2013-2317 PMID:24709929

Toran Pour, A., Moridpour, S., Tay, R., & Rajabifard, A. (2017). Modelling pedestrian crash severity at mid-blocks. Transportmetrica A. *Transportation Science, 13*(3), 273–297.

Toran Pour, A., Moridpour, S., Tay, R., & Rajabifard, A. (Eds.). (2016). A Partial Proportional Odds Model for Pedestrian Crashes at Mid-Blocks in Melbourne Metropolitan Area. In *MATEC Web of Conferences*. EDP Sciences. 10.1051/matecconf/20168102020

Zhu, X., & Lee, C. (2008). Walkability and Safety Around Elementary Schools: Economic and Ethnic Disparities. *American Journal of Preventive Medicine, 34*(4), 282–290. doi:10.1016/j.amepre.2008.01.024 PMID:18374241

Compilation of References

Abdel-Aty, M., Chundi, S. S., & Lee, C. (2007). Geo-spatial and log-linear analysis of pedestrian and bicyclist crashes involving school-aged children. *Journal of Safety Research, 38*(5), 571–579. doi:10.1016/j.jsr.2007.04.006 PMID:18023642

Abdelwahab, H., & Abdel-Aty, M. (2001). Development of artificial neural network models to predict driver injury severity in traffic accidents at signalized intersections. *Transportation Research Record: Journal of the Transportation Research Board*, 6–13.

Abellán, J., López, G., & de Oña, J. (2013). Analysis of traffic accident severity using Decision Rules via Decision Trees. *Expert Systems with Applications, 40*(15), 6047–6054. doi:10.1016/j.eswa.2013.05.027

ABS. (2011). *Australian Bureau of Statistics*. Canberra, Australia: Australia, Year Book.

ABS. (2011b). *Australian Bureau of Statistics*. Available: www.abs.gov.au/

ABS. (2013). *Australia's population by country of birth Canberra*. Australian Bureau of Statistics.

ABS. (2013). *Australia's population by country of birth*. Australian Bureau of Statistics.

ABS. (2015). *National Health Survey: First Results, 2014-15*. Canberra, Australia: Australian Bureau of Statistics 2015 2015. Contract No.: No. 4364.0.55.001.

ABS. (2015). *National Health Survey: First Results, 2014-15*. Canberra, Australia: Australian Bureau of Statistics 2015. Contract No.: No. 4364.0.55.001.

Adeli, H., & Balasubramanyam, K. V. (1988). A synergic man-machine approach to shape optimization of structures. *Computers & Structures, 30*(3), 553–561. doi:10.1016/0045-7949(88)90289-1

Aghabayk, K., Forouzideh, N., & Young, W. (2013). Exploring a Local Linear Model Tree Approach to Car-Following. *Computer-Aided Civil and Infrastructure Engineering, 28*(8), 581–593. doi:10.1111/mice.12011

Agran, P. F., Winn, D. G., Anderson, C. L., & Del Valle, C. (1998). Family, social, and cultural factors in pedestrian injuries among Hispanic children. *Injury Prevention, 4*(3), 188–193. doi:10.1136/ip.4.3.188 PMID:9788088

Agresti, A., & Kateri, M. (2011). *Categorical data analysis*. Springer.

Aguero-Valverde, J., & Jovanis, P. P. (2006). Spatial analysis of fatal and injury crashes in Pennsylvania. *Accident; Analysis and Prevention*, *38*(3), 618–625. doi:10.1016/j.aap.2005.12.006 PMID:16451795

Aho, K., Derryberry, D., & Peterson, T. (2014). Model selection for ecologists: The worldviews of AIC and BIC. *Ecology*, *95*(3), 631–636. doi:10.1890/13-1452.1 PMID:24804445

AIHW. (2010). *Drinking patterns in Australia, 2001–2007*. Canberra: Australia: Australian Institute of Health and Welfare. Contract No.: Cat. no. PHE 133.

Akar, G., Fischer, N., & Namgung, M. (2013). Bicycling choice and gender case study: The Ohio State University. *International Journal of Sustainable Transportation*, *7*(5), 347–365. doi:10.1080/15568318.2012.673694

Al-Ghamdi, A. S. (2002). Pedestrian–vehicle crashes and analytical techniques for stratified contingency tables. *Accident; Analysis and Prevention*, *34*(2), 205–214. doi:10.1016/S0001-4575(01)00015-X PMID:11829290

Al-Shammari, N., Bendak, S., & Al-Gadhi, S. (2009). In-Depth Analysis of Pedestrian Crashes in Riyadh. *Traffic Injury Prevention*, *10*(6), 552–559. doi:10.1080/15389580903175313 PMID:19916125

Altman, J., & Hinkson, M. (2007). Mobility and Modernity in Arnhem Land The Social Universe of Kuninjku Trucks. *Journal of Material Culture*, *12*(2), 181–203. doi:10.1177/1359183507078122

Amoh-Gyimah, R., Sarvi, M., & Saberi, M. (Eds.). (2016). Investigating the Effects of Traffic, Socioeconomic, and Land Use Characteristics on Pedestrian and Bicycle Crashes: A Case Study of Melbourne, Australia. *Transportation Research Board 95th Annual Meeting*.

Anderson, R. W. G., McLean, A. J., Farmer, M. J. B., Lee, B. H., & Brooks, C. G. (1997). Vehicle travel speeds and the incidence of fatal pedestrian crashes. *Accident; Analysis and Prevention*, *29*(5), 667–674. doi:10.1016/S0001-4575(97)00036-5 PMID:9316714

Andrey, J., & Yagar, S. (1993). A temporal analysis of rain-related crash risk. *Accident; Analysis and Prevention*, *25*(4), 465–472. doi:10.1016/0001-4575(93)90076-9 PMID:8357460

Appel, H., Stuertz, G., & Gotzen, L. (1975). Influence of impact speed and vehicle parameter on injuries of children and adults in pedestrian accidents. *Proceedings of the International Research Council on the Biomechanics of Injury conference*, *3*, 83-100.

Assailly, J.-P. (1997). Characterization and prevention of child pedestrian accidents: An overview. *Journal of Applied Developmental Psychology*, *18*(2), 257–262. doi:10.1016/S0193-3973(97)90039-3

Aultman-Hall, L., Lane, D., & Lambert, R. (2009). Assessing Impact of Weather and Season on Pedestrian Traffic Volumes. *Transportation Research Record: Journal of the Transportation Research Board*, *2140*(1), 35–43. doi:10.3141/2140-04

Austroads. (2011). *Application of accessibility measures, AP-R397-11*. Sydney: Austroads.

Aziz, H. M. A., Ukkusuri, S. V., & Hasan, S. (2013). Exploring the determinants of pedestrian–vehicle crash severity in New York City. *Accident; Analysis and Prevention*, *50*(0), 1298–1309. doi:10.1016/j.aap.2012.09.034 PMID:23122781

Ballesteros, M. F., Dischinger, P. C., & Langenberg, P. (2004). Pedestrian injuries and vehicle type in Maryland, 1995–1999. *Accident; Analysis and Prevention*, *36*(1), 73–81. doi:10.1016/S0001-4575(02)00129-X PMID:14572829

Bauman, A. E., Reis, R. S., Sallis, J. F., Wells, J. C., Loos, R. J., Martin, B. W., & Group, L. P. A. S. W. (2012). Correlates of physical activity: Why are some people physically active and others not? *Lancet*, *380*(9838), 258–271. doi:10.1016/S0140-6736(12)60735-1 PMID:22818938

Ben-David, A. (2008). About the relationship between ROC curves and Cohen's kappa. *Engineering Applications of Artificial Intelligence*, *21*(6), 874–882. doi:10.1016/j.engappai.2007.09.009

Bennet, S. A., & Yiannakoulias, N. (2015). Motor-vehicle collisions involving child pedestrians at intersection and mid-block locations. *Accident; Analysis and Prevention*, *78*, 94–103. doi:10.1016/j.aap.2015.03.001 PMID:25756845

Bentley, R., Jolley, D., & Kavanagh, A. M. (2010). Local environments as determinants of walking in Melbourne, Australia. *Social Science & Medicine*, *70*(11), 1806–1815. doi:10.1016/j.socscimed.2010.01.041 PMID:20299141

Bíl, M., Andrášik, R., & Janoška, Z. (2013). Identification of hazardous road locations of traffic accidents by means of kernel density estimation and cluster significance evaluation. *Accident; Analysis and Prevention*, *55*(0), 265–273. doi:10.1016/j.aap.2013.03.003 PMID:23567216

Black, W. R. (1991). Highway accidents: A spatial and temporal analysis. *Transportation Research Record: Journal of the Transportation Research Board*, (1318), 75–82.

Blazquez, C. A., & Celis, M. S. (2013). A spatial and temporal analysis of child pedestrian crashes in Santiago, Chile. *Accident; Analysis and Prevention*, *50*, 304–311. doi:10.1016/j.aap.2012.05.001 PMID:22658462

Boisbunon, A., Canu, S., Fourdrinier, D., Strawderman, W., & Wells, M. T. (2014). Akaike's information criterion, Cp and estimators of loss for elliptically symmetric distributions. *International Statistical Review*, *82*(3), 422–439. doi:10.1111/insr.12052

Bonham, J., & Wilson, A. (2012). Bicycling and the life course: The start-stop-start experiences of women cycling. *International Journal of Sustainable Transportation*, *6*(4), 195–213. doi:10.1080/15568318.2011.585219

Borrell, C., Plasència, A., Huisman, M., Costa, G., Kunst, A., & Andersen, O. (2005). Education level inequalities and transportation injury mortality in the middle aged and elderly in European settings. *Injury Prevention*, *11*(3), 138–142. doi:10.1136/ip.2004.006346 PMID:15933403

Breiman, L., Friedman, J., Stone, C. J., & Olshen, R. A. (1984). *Classification and regression trees*. New York: CRC Press.

Brewer, C. A., & Pickle, L. (2002). Evaluation of methods for classifying epidemiological data on choropleth maps in series. *Annals of the Association of American Geographers*, *92*(4), 662–681. doi:10.1111/1467-8306.00310

Campos-Outcalt, D., Bay, C., Dellapenna, A., & Cota, M. K. (2002). Pedestrian fatalities by race/ethnicity in Arizona, 1990–1996. *American Journal of Preventive Medicine*, *23*(2), 129–135. doi:10.1016/S0749-3797(02)00465-8 PMID:12121801

Cao, X., Handy, S. L., & Mokhtarian, P. L. (2006). The influences of the built environment and residential self-selection on pedestrian behavior: Evidence from Austin, TX. *Transportation*, *33*(1), 1–20. doi:10.100711116-005-7027-2

Carr, L. J., Dunsiger, S. I., & Marcus, B. H. (2010). Walk score™ as a global estimate of neighborhood walkability. *American Journal of Preventive Medicine*, *39*(5), 460–463. doi:10.1016/j.amepre.2010.07.007 PMID:20965384

Ceder, A., Net, Y., & Coriat, C. (2009). Measuring public transport connectivity performance applied in Auckland, New Zealand. *Transportation Research Record: Journal of the Transportation Research Board*, *2111*(1), 139–147. doi:10.3141/2111-16

Celikoglu, H. B. (2013). An Approach to Dynamic Classification of Traffic Flow Patterns. *Computer-Aided Civil and Infrastructure Engineering*, *28*(4), 273–288. doi:10.1111/j.1467-8667.2012.00792.x

Cervero, R., & Duncan, M. (2003). Walking, bicycling, and urban landscapes: Evidence from the San Francisco Bay Area. *American Journal of Public Health*, *93*(9), 1478–1483. doi:10.2105/AJPH.93.9.1478 PMID:12948966

Cervero, R., Sarmiento, O. L., Jacoby, E., Gomez, L. F., & Neiman, A. (2009). Influences of built environments on walking and cycling: Lessons from Bogotá. *International Journal of Sustainable Transportation*, *3*(4), 203–226. doi:10.1080/15568310802178314

Chainey, S., Reid, S., & Stuart, N. (2002). *When is a hotspot a hotspot? A procedure for creating statistically robust hotspot maps of crime*. London: Taylor & Francis.

Chang, L.-Y., & Chien, J.-T. (2013). Analysis of driver injury severity in truck-involved accidents using a non-parametric classification tree model. *Safety Science*, *51*(1), 17–22. doi:10.1016/j.ssci.2012.06.017

Chang, L.-Y., & Wang, H.-W. (2006). Analysis of traffic injury severity: An application of non-parametric classification tree techniques. *Accident; Analysis and Prevention*, *38*(5), 1019–1027. doi:10.1016/j.aap.2006.04.009 PMID:16735022

Cheng, J., & Bertolini, L. (2013). Measuring urban job accessibility with distance decay, competition and diversity. *Journal of Transport Geography*, *30*, 100–109. doi:10.1016/j.jtrangeo.2013.03.005

Cheng, Y.-H., & Chen, S.-Y. (2015). Perceived accessibility, mobility, and connectivity of public transportation systems. *Transportation Research Part A, Policy and Practice*, *77*, 386–403. doi:10.1016/j.tra.2015.05.003

Chen, S., Claramunt, C., & Ray, C. (2014). A spatio-temporal modelling approach for the study of the connectivity and accessibility of the Guangzhou metropolitan network. *Journal of Transport Geography*, *36*, 12–23. doi:10.1016/j.jtrangeo.2014.02.006

Chong, M. M., Abraham, A., & Paprzycki, M. (2005). Traffic Accident Analysis Using Machine Learning Paradigms. *Informatica (Slovenia)*, *29*(1), 89–98.

Chung, Y.-S. (2013). Factor complexity of crash occurrence: An empirical demonstration using boosted regression trees. *Accident; Analysis and Prevention*, *61*, 107–118. doi:10.1016/j.aap.2012.08.015 PMID:22975365

Clifton, K. J., Burnier, C. V., & Akar, G. (2009). Severity of injury resulting from pedestrian–vehicle crashes: What can we learn from examining the built environment? *Transportation Research Part D, Transport and Environment*, *14*(6), 425–436. doi:10.1016/j.trd.2009.01.001

Clifton, K. J., & Kreamer-Fults, K. (2007). An examination of the environmental attributes associated with pedestrian–vehicular crashes near public schools. *Accident; Analysis and Prevention*, *39*(4), 708–715. doi:10.1016/j.aap.2006.11.003 PMID:17174259

Coruh, E., Bilgic, A., & Tortum, A. (2015). Accident analysis with aggregated data: The random parameters negative binomial panel count data model. *Analytic Methods in Accident Research*, *7*, 37–49. doi:10.1016/j.amar.2015.07.001

Cottrill, C. D., & Thakuriah, P. (2010). Evaluating pedestrian crashes in areas with high low-income or minority populations. *Accident; Analysis and Prevention*, *42*(6), 1718–1728. doi:10.1016/j.aap.2010.04.012 PMID:20728622

Coughenour, C., Clark, S., Singh, A., Claw, E., Abelar, J., & Huebner, J. (2017). Examining racial bias as a potential factor in pedestrian crashes. *Accident; Analysis and Prevention*, *98*, 96–100. doi:10.1016/j.aap.2016.09.031 PMID:27716495

Council, M. C. (2008). *Melbourne Bicycle Account: Cycling Census 2008*. City of Melbourne.

Council, M. C. (2016). *Bicycle Plan 2016-2020*. City of Melbourne.

Cui, Z., & Nambisan, S. (2003). Methodology for Evaluating the Safety of Midblock Pedestrian Crossings. *Transportation Research Record: Journal of the Transportation Research Board, 1828*(1), 75-82.

Currie, G., & Stanley, J. (2007). *No way to go: Transport and social disadvantage in Australian communities*. Academic Press.

Currie, G. (2010). Quantifying spatial gaps in public transport supply based on social needs. *Journal of Transport Geography*, *18*(1), 31–41. doi:10.1016/j.jtrangeo.2008.12.002

Daly, A. (1997). Improved methods for trip generation. In *Proceedings of seminar F held at PTRC European Transport Forum*. Brunel University.

Damant-Sirois, G., & El-Geneidy, A. M. (2015). Who cycles more? Determining cycling frequency through a segmentation approach in Montreal, Canada. *Transportation Research Part A, Policy and Practice*, *77*, 113–125. doi:10.1016/j.tra.2015.03.028

Dannenberg, A. L., Jackson, R. J., Frumkin, H., Schieber, R. A., Pratt, M., Kochtitzky, C., & Tilson, H. H. (2003). The impact of community design and land-use choices on public health: A scientific research agenda. *American Journal of Public Health*, *93*(9), 1500–1508. doi:10.2105/AJPH.93.9.1500 PMID:12948970

Data.Vic. (2015). *Victorian Government open data sources*. Retrieved 2015, 2015, from https://www.data.vic.gov.au/

Davis, G. (2001). Relating Severity of Pedestrian Injury to Impact Speed in Vehicle-Pedestrian Crashes: Simple Threshold Model. *Transportation Research Record: Journal of the Transportation Research Board, 1773*(1), 108-13.

De Oña, J., López, G., & Abellán, J. (2013). Extracting decision rules from police accident reports through decision trees. *Accident; Analysis and Prevention*, *50*, 1151–1160. doi:10.1016/j.aap.2012.09.006 PMID:23021419

Dean, J., & Wexler, J. (2014). *What's New in SAS® Enterprise Miner™ 13.1*. Academic Press.

Delmelle, E. C., & Casas, I. (2012). Evaluating the spatial equity of bus rapid transit-based accessibility patterns in a developing country: The case of Cali, Colombia. *Transport Policy*, *20*, 36–46. doi:10.1016/j.tranpol.2011.12.001

Department of Premier and Cabinet. (2015). *Victorian Government Data Directory*. Retrieved from www.data.vic.gov.au

Devkota, B., Dudycha, D., & Andrey, J. (2012). Planning for non-motorised travel in rural Nepal: A role for geographic information systems. *Journal of Transport Geography*, *24*, 282–291. doi:10.1016/j.jtrangeo.2012.03.007

DfT. (n.d.). *Accessibility planning guidance*. Transport for London, Department of Transport.

DiMaggio, C., Mooney, S., Frangos, S., & Wall, S. (2016). Spatial analysis of the association of alcohol outlets and alcohol-related pedestrian/bicyclist injuries in New York City. *Injury Epidemiology*, *3*(1), 1. doi:10.118640621-016-0076-5 PMID:27747548

Dissanayake, S., & Lu, J. J. (2002). Factors influential in making an injury severity difference to older drivers involved in fixed object–passenger car crashes. *Accident; Analysis and Prevention*, *34*(5), 609–618. doi:10.1016/S0001-4575(01)00060-4 PMID:12214955

Dong, H., Ma, L., & Broach, J. (2016). Promoting sustainable travel modes for commute tours: A comparison of the effects of home and work locations and employer-provided incentives. *International Journal of Sustainable Transportation, 10*(6), 485–494. doi:10.1080/15568318.2014.1002027

Dougherty, G., Pless, I. B., & Wilkins, R. (1990). Social class and the occurrence of traffic injuries and deaths in urban children. *Canadian Journal of Public Health, 81*(3), 204–209. PMID:2361207

Duncan, D. T., Aldstadt, J., Whalen, J., Melly, S. J., & Gortmaker, S. L. (2011). Validation of Walk Score® for estimating neighborhood walkability: An analysis of four US metropolitan areas. *International Journal of Environmental Research and Public Health, 8*(11), 4160–4179. doi:10.3390/ijerph8114160 PMID:22163200

EDJTR. (2009). PALI Transport, *Economic Development, Jobs, Transport and Resources.* Victorian Integrated Survey of Travel and Activity (VISTA).

El-Geneidy, A. M., Krizek, K. J., & Iacono, M. (2007). Predicting bicycle travel speeds along different facilities using GPS data: a proof of concept model. *Proceedings of the 86th Annual Meeting of the Transportation Research Board, Compendium of Papers.*

Elias, W., & Shiftan, Y. (2012). The influence of individual's risk perception and attitudes on travel behavior. *Transportation Research Part A, Policy and Practice, 46*(8), 1241–1251. doi:10.1016/j.tra.2012.05.013

Elith, J., Leathwick, J. R., & Hastie, T. (2008). A working guide to boosted regression trees. *Journal of Animal Ecology, 77*(4), 802–813. doi:10.1111/j.1365-2656.2008.01390.x PMID:18397250

Eluru, N., Bhat, C. R., & Hensher, D. A. (2008). A mixed generalized ordered response model for examining pedestrian and bicyclist injury severity level in traffic crashes. *Accident; Analysis and Prevention, 40*(3), 1033–1054. doi:10.1016/j.aap.2007.11.010 PMID:18460372

Elvik, R. (2009). The non-linearity of risk and the promotion of environmentally sustainable transport. *Accident; Analysis and Prevention, 41*(4), 849–855. doi:10.1016/j.aap.2009.04.009 PMID:19540975

Espada, I. & Luk, J. (2011). *Application of accessibility measures.* Academic Press.

Espada, I., & Luk, J. (2011). *Application of accessibility measures.* Academic Press.

Espada, I., Bennett, P., Green, D., & Hatch, D. (2015). *Development of the accessibility-based network operations planning framework.* Academic Press.

Ewing, R., & Cervero, R. (2010). Travel and the built environment. *Journal of the American Planning Association, 76*(3), 265–294. doi:10.1080/01944361003766766

Ewing, R., Schmid, T., Killingsworth, R., Zlot, A., & Raudenbush, S. (2003). Relationship between urban sprawl and physical activity, obesity, and morbidity. *American Journal of Health Promotion, 18*(1), 47–57. doi:10.4278/0890-1171-18.1.47 PMID:13677962

Factor, R., Mahalel, D., & Yair, G. (2007). The social accident: A theoretical model and a research agenda for studying the influence of social and cultural characteristics on motor vehicle accidents. *Accident; Analysis and Prevention*, *39*(5), 914–921. doi:10.1016/j.aap.2006.12.015 PMID:17291438

Fernández-Heredia, Á., Monzón, A., & Jara-Díaz, S. (2014). Understanding cyclists' perceptions, keys for a successful bicycle promotion. *Transportation Research Part A, Policy and Practice*, *63*, 1–11. doi:10.1016/j.tra.2014.02.013

Fox, L., Serre, M. L., Lippmann, S. J., Rodríguez, D. A., Bangdiwala, S. I., Gutiérrez, M. I., ... Villaveces, A. (2015). Spatiotemporal Approaches to Analyzing Pedestrian Fatalities: The Case of Cali, Colombia. *Traffic Injury Prevention*, *16*(6), 571–577. doi:10.1080/15389588.2014.976336 PMID:25551356

Frank, L. D., Andresen, M. A., & Schmid, T. L. (2004). Obesity relationships with community design, physical activity, and time spent in cars. *American Journal of Preventive Medicine*, *27*(2), 87–96. doi:10.1016/j.amepre.2004.04.011 PMID:15261894

Frank, L. D., Sallis, J. F., Conway, T. L., Chapman, J. E., Saelens, B. E., & Bachman, W. (2006). Many pathways from land use to health: Associations between neighborhood walkability and active transportation, body mass index, and air quality. *Journal of the American Planning Association*, *72*(1), 75–87. doi:10.1080/01944360608976725

Frank, L. D., Sallis, J. F., Saelens, B. E., Leary, L., Cain, K., Conway, T. L., & Hess, P. M. (2010). The development of a walkability index: Application to the Neighborhood Quality of Life Study. *British Journal of Sports Medicine*, *44*(13), 924–933. doi:10.1136/bjsm.2009.058701 PMID:19406732

Frank, L. D., Schmid, T. L., Sallis, J. F., Chapman, J., & Saelens, B. E. (2005). Linking objectively measured physical activity with objectively measured urban form: Findings from SMARTRAQ. *American Journal of Preventive Medicine*, *28*(2), 117–125. doi:10.1016/j.amepre.2004.11.001 PMID:15694519

Frank, L., Engelke, P., & Schmid, T. (2003). *Health and community design: The impact of the built environment on physical activity*. Island Press.

Fransen, K., Neutens, T., Farber, S., De Maeyer, P., Deruyter, G., & Witlox, F. (2015). Identifying public transport gaps using time-dependent accessibility levels. *Journal of Transport Geography*, *48*, 176–187. doi:10.1016/j.jtrangeo.2015.09.008

Friedman, J., Hastie, T., & Tibshirani, R. (2001). The elements of statistical learning. Springer.

Friedman, J. H., & Meulman, J. J. (2003). Multiple additive regression trees with application in epidemiology. *Statistics in Medicine*, *22*(9), 1365–1381. doi:10.1002im.1501 PMID:12704603

Galanis, A., Papanikolaou, A., & Eliou, N. (2014). Bikeability Audit in Urban Road Environment: Case Study in the City of Volos, Greece. *International Journal of Operations Research and Information Systems*, *5*(2), 21–39. doi:10.4018/ijoris.2014040102

Geurs, K. T., & Van Wee, B. (2004). Accessibility evaluation of land-use and transport strategies: Review and research directions. *Journal of Transport Geography*, *12*(2), 127–140. doi:10.1016/j.jtrangeo.2003.10.005

Giles-Corti, B., Macaulay, G., Middleton, N., Boruff, B., Bull, F., Butterworth, I., ... Christian, H. (2015). Developing a research and practice tool to measure walkability: A demonstration project. *Health Promotion Journal of Australia*, *25*(3), 160–166. doi:10.1071/HE14050 PMID:25481614

Gómez, L. F., Parra, D. C., Buchner, D., Brownson, R. C., Sarmiento, O. L., Pinzón, J. D., ... Lobelo, F. (2010). Built Environment Attributes and Walking Patterns Among the Elderly Population in Bogotá. *American Journal of Preventive Medicine*, *38*(6), 592–599. doi:10.1016/j.amepre.2010.02.005 PMID:20494235

Goodman, L. A., & Kruskal, W. H. (1954). Measures of association for cross classifications. *Journal of the American Statistical Association*, *49*, 732–764.

Graham, D. J., & Glaister, S. (2003). Spatial variation in road pedestrian casualties: The role of urban scale, density and land-use mix. *Urban Studies (Edinburgh, Scotland)*, *40*(8), 1591–1607. doi:10.1080/0042098032000094441

Graham, D., Glaister, S., & Anderson, R. (2005). The effects of area deprivation on the incidence of child and adult pedestrian casualties in England. *Accident; Analysis and Prevention*, *37*(1), 125–135. doi:10.1016/j.aap.2004.07.002 PMID:15607283

Handy, S. L., & Clifton, K. J. (2001). Evaluating neighborhood accessibility: Possibilities and practicalities. *Journal of Transportation and Statistics*, *4*(2/3), 67–78.

Handy, S. L., & Xing, Y. (2011). Factors correlated with bicycle commuting: A study in six small US cities. *International Journal of Sustainable Transportation*, *5*(2), 91–110. doi:10.1080/15568310903514789

Hanneman, R. A., Kposowa, A. J., & Riddle, M. D. (2012). *Basic Statistics for Social Research*. John Wiley & Sons.

Hansen, W. G. (1959). How accessibility shapes land use. *Journal of the American Institute of Planners*, *25*(2), 73–76. doi:10.1080/01944365908978307

Harwood, D. W., Bauer, K. M., Richard, K. R., Gilmore, D. K., Graham, J. L., Potts, I. B., & ... . (2008). *Pedestrian Safety Prediction Methodology. NCHRP Web-only Document 129: Phase III*. Washington, DC: Transportation Research Board.

Hassan, H. M., Shawky, M., Kishta, M., Garib, A. M., & Al-Harthei, H. A. (2017). Investigation of drivers' behavior towards speeds using crash data and self-reported questionnaire. *Accident; Analysis and Prevention*, *98*, 348–358. doi:10.1016/j.aap.2016.10.027 PMID:27837722

Hayashi, T., Tsumura, K., Suematsu, C., Okada, K., Fujii, S., & Endo, G. (1999). Walking to work and the risk for hypertension in men: The Osaka Health Survey. *Annals of Internal Medicine*, *131*(1), 21–26. doi:10.7326/0003-4819-131-1-199907060-00005 PMID:10391811

Henary, B. Y., Ivarsson, J., & Crandall, J. R. (2006). The Influence of Age on the Morbidity and Mortality of Pedestrian Victims. *Traffic Injury Prevention*, *7*(2), 182–190. doi:10.1080/15389580500516414 PMID:16854713

Herabat, P., & Songchitruksa, P. (2003). A Decision Support System for Flexible Pavement Treatment Selection. *Computer-Aided Civil and Infrastructure Engineering*, *18*(2), 147–160. doi:10.1111/1467-8667.00306

Hilbe, J. M. (2008). *Brief overview on interpreting count model risk ratios: An addendum to negative binomial regression*. Cambridge University Press.

Hilbe, J. M. (2011). *Negative binomial regression*. Cambridge University Press. doi:10.1017/CBO9780511973420

Holland, C., & Hill, R. (2007). The effect of age, gender and driver status on pedestrians' intentions to cross the road in risky situations. *Accident; Analysis and Prevention*, *39*(2), 224–237. doi:10.1016/j.aap.2006.07.003 PMID:16979132

Holubowycz, O. T. (1995). Age, sex, and blood alcohol concentration of killed and injured pedestrians. *Accident; Analysis and Prevention*, *27*(3), 417–422. doi:10.1016/0001-4575(94)00064-S PMID:7639925

Hoogendoorn, S. P., Bovy, P., & Daamen, W. (2004). Walking infrastructure design assessment by continuous space dynamic assignment modeling. *Journal of Advanced Transportation*, *38*(1), 69–92. doi:10.1002/atr.5670380106

Hosseinpour, M., Prasetijo, J., Yahaya, A. S., & Ghadiri, S. M. R. (2013). A Comparative Study of Count Models: Application to Pedestrian-Vehicle Crashes Along Malaysia Federal Roads. *Traffic Injury Prevention*, *14*(6), 630–638. doi:10.1080/15389588.2012.736649 PMID:23859313

Hung, S.-L., & Jan, J. C. (1999). Machine Learning in Engineering Analysis and Design: An Integrated Fuzzy Neural Network Learning Model. *Computer-Aided Civil and Infrastructure Engineering*, *14*(3), 207–219. doi:10.1111/0885-9507.00142

Hupkes, G. (1982). The law of constant travel time and trip-rates. *Futures*, *14*(1), 38–46. doi:10.1016/0016-3287(82)90070-2

Hurni, A. (2005). *Transport and social exclusion in Western Sydney*. Australian Transport Research Forum (ATRF), 28th, 2005, Sydney, New South Wales, Austalia.

Hu, S. (2007). *Akaike information criterion*. Center for Research in Scientific Computation.

Iacono, M., Krizek, K. J., & El-Geneidy, A. (2010). Measuring non-motorised accessibility: Issues, alternatives, and execution. *Journal of Transport Geography*, *18*(1), 133–140. doi:10.1016/j.jtrangeo.2009.02.002

Ian, H. (2010). *An introduction to geographical information systems*. Pearson Education India.

Ishaque, M. M., & Noland, R. B. (2008). Behavioural Issues in Pedestrian Speed Choice and Street Crossing Behaviour: A Review. *Transport Reviews*, *28*(1), 61–85. doi:10.1080/01441640701365239

Jiang, X., Abdel-Aty, M., Hu, J., & Lee, J. (2016). Investigating macro-level hotzone identification and variable importance using big data: A random forest models approach. *Neurocomputing*, *181*, 53–63. doi:10.1016/j.neucom.2015.08.097

Johnson, V., Currie, G., & Stanley, J. (2011). Exploring transport to arts and cultural activities as a facilitator of social inclusion. *Transport Policy*, *18*(1), 68–75. doi:10.1016/j.tranpol.2010.06.001

Jun, M.-J., Kim, J. I., Kwon, J. H., & Jeong, J.-E. (2012). The effects of high-density suburban development on commuter mode choices in Seoul, Korea. *Cities (London, England)*.

Kashani, A. T., & Mohaymany, A. S. (2011). Analysis of the traffic injury severity on two-lane, two-way rural roads based on classification tree models. *Safety Science*, *49*(10), 1314–1320. doi:10.1016/j.ssci.2011.04.019

Kerrigan, M., & Bull, D. (1992). Measuring accessibility: A public transport accessibility index. *Environmental Issues. Proceeding of 20th Annual Seminar B Held at the PTRC European Transport, 354.*

Kim, J.-K., Ulfarsson, G. F., Shankar, V. N., & Kim, S. (2008). Age and pedestrian injury severity in motor-vehicle crashes: A heteroskedastic logit analysis. *Accident; Analysis and Prevention*, *40*(5), 1695–1702. doi:10.1016/j.aap.2008.06.005 PMID:18760098

Kim, J.-K., Ulfarsson, G. F., Shankar, V. N., & Mannering, F. L. (2010a). A note on modeling pedestrian-injury severity in motor-vehicle crashes with the mixed logit model. *Accident; Analysis and Prevention*, *42*(6), 1751–1758. doi:10.1016/j.aap.2010.04.016 PMID:20728626

Y.-I. Kim, S. H. Park, & S.-Y. Kho (Eds.). (2010b). *A Case Study of 'Continuous Risk Profile' Approach for Hotspots Identification on Korean Expressways. 17th ITS World Congress*, Busan, South Korea.

Koepsell, T., McCloskey, L., & Wolf, M. (2002). CRosswalk markings and the risk of pedestrian–motor vehicle collisions in older pedestrians. *Journal of the American Medical Association*, *288*(17), 2136–2143. doi:10.1001/jama.288.17.2136 PMID:12413373

Koopmans, J. M., Friedman, L., Kwon, S., & Sheehan, K. (2015). Urban crash-related child pedestrian injury incidence and characteristics associated with injury severity. *Accident; Analysis and Prevention*, *77*, 127–136. doi:10.1016/j.aap.2015.02.005 PMID:25703351

Krizek, K. J. (2005). *Perspectives on accessibility and travel*. Academic Press.

Kuhnert, P. M., Do, K.-A., & McClure, R. (2000). Combining non-parametric models with logistic regression: An application to motor vehicle injury data. *Computational Statistics & Data Analysis*, *34*(3), 371–386. doi:10.1016/S0167-9473(99)00099-7

Kuhn, M. (2008). Caret package. *Journal of Statistical Software*, *28*(5).

Kwan, M. P. (1999). Gender and individual access to urban opportunities: A study using space–time measures. *The Professional Geographer*, *51*(2), 210–227. doi:10.1111/0033-0124.00158

Kwan, M.-P. (1998). Space-time and integral measures of individual accessibility: A comparative analysis using a point-based framework. *Geographical Analysis*, *30*(3), 191–216. doi:10.1111/j.1538-4632.1998.tb00396.x

Kwan, M.-P. (2013). Beyond Space (As We Knew It): Toward Temporally Integrated Geographies of Segregation, Health, and Accessibility: Space–Time Integration in Geography and GIScience. *Annals of the Association of American Geographers*, *103*(5), 1078–1086. doi:10.1080/00045608.2013.792177

Kwan, M.-P. (2015). *Beyond Space (As We Knew It): Toward Temporally Integrated Geographies of Segregation, Health, and Accessibility. In Space-Time Integration in Geography and GIScience*. Springer.

Kwan, M.-P., Murray, A. T., O'Kelly, M. E., & Tiefelsdorf, M. (2003). Recent advances in accessibility research: Representation, methodology and applications. *Journal of Geographical Systems*, *5*(1), 129–138. doi:10.1007101090300107

Kwon, O. H., Rhee, W., & Yoon, Y. (2015). Application of classification algorithms for analysis of road safety risk factor dependencies. *Accident; Analysis and Prevention*, *75*, 1–15. doi:10.1016/j.aap.2014.11.005 PMID:25460086

Laflamme, L., Vaez, M., Hasselberg, M., & Kullgren, A. (2005). Car safety and social differences in traffic injuries among young adult drivers: A study of two-car injury-generating crashes in Sweden. *Safety Science*, *43*(1), 1–10. doi:10.1016/j.ssci.2004.09.001

LaScala, E. A., Gerber, D., & Gruenewald, P. J. (2000). Demographic and environmental correlates of pedestrian injury collisions: A spatial analysis. *Accident; Analysis and Prevention*, *32*(5), 651–658. doi:10.1016/S0001-4575(99)00100-1 PMID:10908137

LaScala, E. A., Gruenewald, P. J., & Johnson, F. W. (2004). An ecological study of the locations of schools and child pedestrian injury collisions. *Accident; Analysis and Prevention*, *36*(4), 569–576. doi:10.1016/S0001-4575(03)00063-0 PMID:15094409

Lavizzo-Mourey, R., & Mcginnis, J. M. (2003). Making the case for active living communities. *American Journal of Public Health*, *93*(9), 1386–1388. doi:10.2105/AJPH.93.9.1386 PMID:12948948

Lee, C., & Abdel-Aty, M. (2005). Comprehensive analysis of vehicle–pedestrian crashes at intersections in Florida. *Accident; Analysis and Prevention*, *37*(4), 775–786. doi:10.1016/j.aap.2005.03.019 PMID:15869737

Lee, G., Park, Y., Kim, J., & Cho, G.-H. (2016). Association between intersection characteristics and perceived crash risk among school-aged children. *Accident; Analysis and Prevention*, *97*, 111–121. doi:10.1016/j.aap.2016.09.001 PMID:27612169

Lee, J.-S., Nam, J., & Lee, S.-S. (2014). Built environment impacts on individual mode choice: An empirical study of the Houston-Galveston metropolitan area. *International Journal of Sustainable Transportation*, *8*(6), 447–470. doi:10.1080/15568318.2012.716142

Lefler, D. E., & Gabler, H. C. (2004). The fatality and injury risk of light truck impacts with pedestrians in the United States. *Accident; Analysis and Prevention*, *36*(2), 295–304. doi:10.1016/S0001-4575(03)00007-1 PMID:14642884

Lei, T., & Church, R. (2010). Mapping transit-based access: Integrating GIS, routes and schedules. *International Journal of Geographical Information Science*, *24*(2), 283–304. doi:10.1080/13658810902835404

Levine, J. (2010). *Zoned out: Regulation, markets, and choices in transportation and metropolitan land use*. Routledge. doi:10.4324/9781936331215

Levine, N., Kim, K. E., & Nitz, L. H. (1995). Spatial analysis of Honolulu motor vehicle crashes: I. Spatial patterns. *Accident; Analysis and Prevention*, *27*(5), 663–674. doi:10.1016/0001-4575(95)00017-T PMID:8579697

Li, D., Ranjitkar, P., Zhao, Y., Yi, H., & Rashidi, S. (2016). Analyzing pedestrian crash injury severity under different weather conditions. *Traffic Injury Prevention*.

Li, L., Zhu, L., & Sui, D. Z. (2007). A GIS-based Bayesian approach for analyzing spatial–temporal patterns of intra-city motor vehicle crashes. *Journal of Transport Geography*, *15*(4), 274–285. doi:10.1016/j.jtrangeo.2006.08.005

London, T. F. (2010). *Measuring Public Transport Accessibility Levels*. Academic Press.

Loo, B. P. Y., & Tsui, M. K. (2005). Temporal and spatial patterns of vehicle-pedestrian crashes in busy commercial and shopping areas: A case study of hong kong. *Asian Geographer*, *24*(1-2), 113–128. doi:10.1080/10225706.2005.9684124

Lo, R. H. (2009). Walkability: What is it? *Journal of Urbanism*, *2*(2), 145–166. doi:10.1080/17549170903092867

Lord, D., van Schalkwyk, I., Chrysler, S., & Staplin, L. (2007). A strategy to reduce older driver injuries at intersections using more accommodating roundabout design practices. *Accident; Analysis and Prevention*, *39*(3), 427–432. doi:10.1016/j.aap.2006.09.011 PMID:17092474

Lowry, M. B., Furth, P., & Hadden-Loh, T. (2016). Prioritizing new bicycle facilities to improve low-stress network connectivity. *Transportation Research Part A, Policy and Practice*, *86*, 124–140. doi:10.1016/j.tra.2016.02.003

Lucas, K. (2011). Making the connections between transport disadvantage and the social exclusion of low income populations in the Tshwane Region of South Africa. *Journal of Transport Geography*, *19*(6), 1320–1334. doi:10.1016/j.jtrangeo.2011.02.007

Lucas, K. (2012). Transport and social exclusion: Where are we now? *Transport Policy*, *20*, 105–113. doi:10.1016/j.tranpol.2012.01.013

Lyons, R. A., Towner, E., Christie, N., Kendrick, D., Jones, S. J., Hayes, M., ... Phillips, C. (2008). The Advocacy in Action Study a cluster randomized controlled trial to reduce pedestrian injuries in deprived communities. *Injury Prevention*, *14*(2), e1. doi:10.1136/ip.2007.017632 PMID:18388222

Manaugh, K., & El-Geneidy, A. (2011). Validating walkability indices: How do different households respond to the walkability of their neighborhood? *Transportation Research Part D, Transport and Environment*, *16*(4), 309–315. doi:10.1016/j.trd.2011.01.009

Manson, J. E., Hu, F. B., Rich-Edwards, J. W., Colditz, G. A., Stampfer, M. J., Willett, W. C., ... Hennekens, C. H. (1999). A prospective study of walking as compared with vigorous exercise in the prevention of coronary heart disease in women. *The New England Journal of Medicine*, *341*(9), 650–658. doi:10.1056/NEJM199908263410904 PMID:10460816

Mavoa, S., Witten, K., Mccreanor, T., & O'Sullivan, D. (2012). GIS based destination accessibility via public transit and walking in Auckland, New Zealand. *Journal of Transport Geography*, *20*(1), 15–22. doi:10.1016/j.jtrangeo.2011.10.001

Mazloumi, E., Moridpour, S., Currie, G., & Rose, G. (2011). Exploring the value of traffic flow data in bus travel time prediction. *Journal of Transportation Engineering*, *138*(4), 436–446. doi:10.1061/(ASCE)TE.1943-5436.0000329

McDonald, N. C. (2007). Active transportation to school: Trends among US schoolchildren, 1969–2001. *American Journal of Preventive Medicine*, *32*(6), 509–516. doi:10.1016/j.amepre.2007.02.022 PMID:17533067

McNeil, N. (2011). Bikeability and the 20-min neighborhood: How infrastructure and destinations influence bicycle accessibility. *Transportation Research Record: Journal of the Transportation Research Board*, *2247*(1), 53–63. doi:10.3141/2247-07

Mekuria, M. C., Furth, P. G., & Nixon, H. (2012). *Low-stress bicycling and network connectivity*. Academic Press.

Mesbah, M., Thompson, R., & Moridpour, S. (2012). Bilevel optimization approach to design of network of bike lanes. *Transportation Research Record: Journal of the Transportation Research Board*, *2284*(1), 21–28. doi:10.3141/2284-03

Milakis, D., Cervero, R., Van Wee, B., & Maat, K. (2015). Do people consider an acceptable travel time? Evidence from Berkeley, CA. *Journal of Transport Geography*, *44*, 76–86. doi:10.1016/j.jtrangeo.2015.03.008

Miles-Doan, R. (1996). Alcohol use among pedestrians and the odds of surviving an injury: Evidence from Florida law enforcement data. *Accident; Analysis and Prevention*, *28*(1), 23–31. doi:10.1016/0001-4575(95)00030-5 PMID:8924182

Millward, H., Spinney, J., & Scott, D. (2013). Active-transport walking behavior: Destinations, durations, distances. *Journal of Transport Geography*, *28*, 101–110. doi:10.1016/j.jtrangeo.2012.11.012

Miranda, S., Carrasco, Y., & Jorge, G. (2011). *Pedestrian Volume Studies: A case study in the city of Gothenburg*. Göteborg, Sweden: Chalmers University of Technology.

Miranda-Moreno, L. F., Morency, P., & El-Geneidy, A. M. (2011). The link between built environment, pedestrian activity and pedestrian–vehicle collision occurrence at signalized intersections. *Accident; Analysis and Prevention*, *43*(5), 1624–1634. doi:10.1016/j.aap.2011.02.005 PMID:21658488

Mitra, R., & Buliung, R. N. (2012). Built environment correlates of active school transportation: Neighborhood and the modifiable areal unit problem. *Journal of Transport Geography*, *20*(1), 51–61. doi:10.1016/j.jtrangeo.2011.07.009

Morency, P., Gauvin, L., Plante, C., Fournier, M., & Morency, C. (2012). Neighborhood Social Inequalities in Road Traffic Injuries: The Influence of Traffic Volume and Road Design. *American Journal of Public Health*, *102*(6), 1112–1119. doi:10.2105/AJPH.2011.300528 PMID:22515869

Morrison, C., Ponicki, W. R., Gruenewald, P. J., Wiebe, D. J., & Smith, K. (2016). Spatial relationships between alcohol-related road crashes and retail alcohol availability. *Drug and Alcohol Dependence*, *162*, 241–244. doi:10.1016/j.drugalcdep.2016.02.033 PMID:26968094

Morse-Mcnabb, E. (2011). The Victorian Land Use Information System (VLUIS): A new method for creating land use data for Victoria, *Australia. Surveying and Spatial Sciences Conference*, 155.

Motoaki, Y., & Daziano, R. A. (2015). A hybrid-choice latent-class model for the analysis of the effects of weather on cycling demand. *Transportation Research Part A, Policy and Practice*, *75*, 217–230. doi:10.1016/j.tra.2015.03.017

Moudon, A. V., Lin, L., Jiao, J., Hurvitz, P., & Reeves, P. (2011). The risk of pedestrian injury and fatality in collisions with motor vehicles, a social ecological study of state routes and city streets in King County, Washington. *Accident; Analysis and Prevention*, *43*(1), 11–24. doi:10.1016/j.aap.2009.12.008 PMID:21094292

Neutens, T., Delafontaine, M., Scott, D. M., & De Maeyer, P. (2012). An analysis of day-to-day variations in individual space–time accessibility. *Journal of Transport Geography*, *23*, 81–91. doi:10.1016/j.jtrangeo.2012.04.001

Neutens, T., Witlox, F., Van De Weghe, N., & De Maeyer, P. (2007). Space–time opportunities for multiple agents: A constraint-based approach. *International Journal of Geographical Information Science*, *21*(10), 1061–1076. doi:10.1080/13658810601169873

Newson, R. (2006). Confidence intervals for rank statistics: Somers' D and extensions. *The Stata Journal*, *6*(3), 309–334. doi:10.1177/1536867X0600600302

Newstead, S., & D'Elia, A. (2010). Does vehicle colour influence crash risk? *Safety Science*, *48*(10), 1327–1338. doi:10.1016/j.ssci.2010.05.001

NHTSA, . (2015). Traffic Safety Facts, 2012 Data: Pedestrians. *Annals of Emergency Medicine*, *65*(4), 452. doi:10.1016/j.annemergmed.2015.02.019

Nicaj, L., Wilt, S., & Henning, K. (2006). Motor vehicle crash pedestrian deaths in New York City: The plight of the older pedestrian. *Injury Prevention, 12*(6), 414–416. doi:10.1136/ip.2005.010082 PMID:17170193

Noland, R. B., & Oh, L. (2004). The effect of infrastructure and demographic change on traffic-related fatalities and crashes: A case study of Illinois county-level data. *Accident; Analysis and Prevention, 36*(4), 525–532. doi:10.1016/S0001-4575(03)00058-7 PMID:15094404

Noland, R., & Quddus, M. (2004). Analysis of pedestrian and bicycle casualties with regional panel data. *Transportation Research Record: Journal of the Transportation Research Board*, 28–33.

O'Sullivan, D., & Unwin, D. J. (2014). *Geographic information analysis*. Hoboken, NJ: John Wiley & Sons.

Oikawa, S., Matsui, Y., Doi, T., & Sakurai, T. (2016). Relation between vehicle travel velocity and pedestrian injury risk in different age groups for the design of a pedestrian detection system. *Safety Science, 82*, 361–367. doi:10.1016/j.ssci.2015.10.003

Ortúzar, J. D., Iacobelli, A., & Valeze, C. (2000). Estimating demand for a cycle-way network. *Transportation Research Part A, Policy and Practice, 34*(5), 353–373. doi:10.1016/S0965-8564(99)00040-3

Owen, N., Cerin, E., Leslie, E., Coffee, N., Frank, L. D., Bauman, A. E., ... Sallis, J. F. (2007a). Neighborhood walkability and the walking behavior of Australian adults. *American Journal of Preventive Medicine, 33*(5), 387–395. doi:10.1016/j.amepre.2007.07.025 PMID:17950404

Paris. (2010). *Moving in Paris, Mairie de Paris*. Paris.

Parkin, J., & Rotheram, J. (2010). Design speeds and acceleration characteristics of bicycle traffic for use in planning, design and appraisal. *Transport Policy, 17*(5), 335–341. doi:10.1016/j.tranpol.2010.03.001

Pasanen, E., & Salmivaara, H. (1993). Driving speeds and pedestrian safety in the City of Helsinki. *Traffic Engineering & Control, 34*(6), 308–310.

Peiravian, F., Derrible, S., & Ijaz, F. (2014). Development and application of the Pedestrian Environment Index (PEI). *Journal of Transport Geography, 39*, 73–84. doi:10.1016/j.jtrangeo.2014.06.020

Peña-garcía, A., De Oña, R., García, P., Peña-garcía, P., & de Oña, J. (2014). Effects of Daytime Running Lamps on Pedestrians Visual Reaction Time: Implications on Vehicles and Human Factors. *Procedia Engineering, 84*, 603–607. doi:10.1016/j.proeng.2014.10.473

Pink, B. (2010). *Year Book Australia*. Contract No.: ABS Catalogue No. 1301.0.

Pink, B. (2011). Australian statistical geography standard (ASGS): volume 5–remoteness structure. Canberra: Australian Bureau of Statistics.

Plug, C., Xia, J. C., & Caulfield, C. (2011). Spatial and temporal visualisation techniques for crash analysis. *Accident; Analysis and Prevention*, *43*(6), 1937–1946. doi:10.1016/j.aap.2011.05.007 PMID:21819821

Priya, T., & Uteng, A. (2009). Dynamics of transport and social exclusion: Effects of expensive driver's license. *Transport Policy*, *16*(3), 130–139. doi:10.1016/j.tranpol.2009.02.005

Prud'homme, R., & Bocarejo, J. P. (2005). The London congestion charge: A tentative economic appraisal. *Transport Policy*, *12*(3), 279–287. doi:10.1016/j.tranpol.2005.03.001

PSMA. (2011). *2012 Annual report*. Canberra, Australia: Mapping Data for Australia.

PSMA. (2011). *Features of Interest*. Data Product Description.

Pulugurtha, S. S., Krishnakumar, V. K., & Nambisan, S. S. (2007). New methods to identify and rank high pedestrian crash zones: An illustration. *Accident; Analysis and Prevention*, *39*(4), 800–811. doi:10.1016/j.aap.2006.12.001 PMID:17227666

Pulugurtha, S. S., & Sambhara, V. R. (2011). Pedestrian crash estimation models for signalized intersections. *Accident; Analysis and Prevention*, *43*(1), 439–446. doi:10.1016/j.aap.2010.09.014 PMID:21094342

Quistberg, D. A., Howard, E. J., Ebel, B. E., Moudon, A. V., Saelens, B. E., Hurvitz, P. M., ... Rivara, F. P. (2015). Multilevel models for evaluating the risk of pedestrian–motor vehicle collisions at intersections and mid-blocks. *Accident; Analysis and Prevention*, *84*, 99–111. doi:10.1016/j.aap.2015.08.013 PMID:26339944

Rahul, T. M., & Verma, A. (2014). A study of acceptable trip distances using walking and cycling in Bangalore. *Journal of Transport Geography*, *38*, 106–113. doi:10.1016/j.jtrangeo.2014.05.011

Rattan, A., Campese, A. & Eden, C. (2012). Modeling walkability. *Arc. User. Winter, 2012*, 30-3.

REDFIN. (2015). *How Walk Scor Works*. Available: https://www.redfin.com

Reich, Y. (1997). Machine Learning Techniques for Civil Engineering Problems. *Computer-Aided Civil and Infrastructure Engineering*, *12*(4), 295–310. doi:10.1111/0885-9507.00065

Rendall, S., Page, S., Reitsma, F., Van Houten, E., & Krumdieck, S. (2011). Quantifying transport energy resilience: Active mode accessibility. *Transportation Research Record: Journal of the Transportation Research Board*, *2242*(1), 72–80. doi:10.3141/2242-09

Rezaei, M. A., Behzadi, G., Ahmadian, S., & Rezaei, M. (2013). Pedestrian's Accidents Prediction in Suburban Roads Using Artificial Neural Network (Case study of Amol city). *Journal of Intelligent Transportation and Urban Planning.*, *1*(1), 41–47. doi:10.18005/ITUP0101005

Ridgeway, G. (2007). Generalized Boosted Models: A guide to the gbm package. *Update.*, *1*(1), 2007.

Rifaat, S.M., Tay, R., Raihan, S.M., Fahim, A., & Touhidduzzaman, S.M. (2017). Vehicle-Pedestrian crashes at Intersections in Dhaka city. *The Open Transportation Journal, 11*(1).

Rifaat, S. M., Tay, R., & de Barros, A. (2012). Urban Street Pattern and Pedestrian Traffic Safety. *Journal of Urban Design*, *17*(3), 337–352. doi:10.1080/13574809.2012.683398

Rothman, L., Macarthur, C., To, T., Buliung, R., & Howard, A. (2014). Motor Vehicle-Pedestrian Collisions and Walking to School: The Role of the Built Environment. *Pediatrics*, *133*(5), 776–784. doi:10.1542/peds.2013-2317 PMID:24709929

Saelens, B. E., Sallis, J. F., Black, J. B., & Chen, D. (2003). Neighborhood-based differences in physical activity: An environment scale evaluation. *American Journal of Public Health*, *93*(9), 1552–1558. doi:10.2105/AJPH.93.9.1552 PMID:12948979

Saghapour, T., Moridpour, S., & Thompson, R. G. (2017). Estimating Walking Access Levels Incorporating Distance Thresholds of Built Environment Features. *International Journal of Sustainable Transportation*, 1–14. doi:10.1080/15568318.2017.1380245

Saha, D., Alluri, P., & Gan, A. (2015). Prioritizing Highway Safety Manual's crash prediction variables using boosted regression trees. *Accident; Analysis and Prevention*, *79*(0), 133–144. doi:10.1016/j.aap.2015.03.011 PMID:25823903

Sarkar, S., Tay, R., & Hunt, J. (2011). Logistic Regression Model of Risk of Fatality in Vehicle-Pedestrian Crashes on National Highways in Bangladesh. *Transportation Research Record: Journal of the Transportation Research Board, 2264*(1), 128-37.

Savolainen, P. T., Mannering, F. L., Lord, D., & Quddus, M. A. (2011). The statistical analysis of highway crash-injury severities: A review and assessment of methodological alternatives. *Accident; Analysis and Prevention*, *43*(5), 1666–1676. doi:10.1016/j.aap.2011.03.025 PMID:21658493

Schneider, R. J., Ryznar, R. M., & Khattak, A. J. (2004). An accident waiting to happen: A spatial approach to proactive pedestrian planning. *Accident; Analysis and Prevention*, *36*(2), 193–211. doi:10.1016/S0001-4575(02)00149-5 PMID:14642874

Schwanen, T., & Dijst, M. (2002). Travel-time ratios for visits to the workplace: The relationship between commuting time and work duration. *Transportation Research Part A, Policy and Practice*, *36*(7), 573–592. doi:10.1016/S0965-8564(01)00023-4

Shay, E., & Khattak, A. J. (2012). Household travel decision chains: Residential environment, automobile ownership, trips and mode choice. *International Journal of Sustainable Transportation*, *6*(2), 88–110. doi:10.1080/15568318.2011.560363

Shinar, D., Schechtman, E., & Compton, R. (2001). Self-reports of safe driving behaviors in relationship to sex, age, education and income in the US adult driving population. *Accident; Analysis and Prevention*, *33*(1), 111–116. doi:10.1016/S0001-4575(00)00021-X PMID:11189114

Siddiqui, C., Abdel-Aty, M., & Choi, K. (2012). Macroscopic spatial analysis of pedestrian and bicycle crashes. *Accident; Analysis and Prevention*, *45*, 382–391. doi:10.1016/j.aap.2011.08.003 PMID:22269522

Silverman, B. W. (1986). *Density estimation for statistics and data analysis*. London: CRC press. doi:10.1007/978-1-4899-3324-9

Sinnott, R., Galang, G., Tomko, M. & Stimson, R. (2011). *Australian Urban Research Infrastructure Network*. Academic Press.

Sorton, A., & Walsh, T. (1994). Bicycle stress level as a tool to evaluate urban and suburban bicycle compatibility. *Transportation Research Record: Journal of the Transportation Research Board*, 17–17.

Sprinthall, R. C. (2011). *Basic statistical analysis*. Academic Press.

Sundquist, K., Eriksson, U., Kawakami, N., Skog, L., Ohlsson, H., & Arvidsson, D. (2011). Neighborhood walkability, physical activity, and walking behavior: The Swedish Neighborhood and Physical Activity (SNAP) study. *Social Science & Medicine*, *72*(8), 1266–1273. doi:10.1016/j.socscimed.2011.03.004 PMID:21470735

Swanson, K. C., & Mccormack, G. R. (2012). The Relations Between Driving Behavior, Physical Activity, and Weight Status Among Canadian Adults. *Journal of Physical Activity & Health*, *9*(3), 352–359. doi:10.1123/jpah.9.3.352 PMID:21934155

Sze, N. N., & Wong, S. C. (2007). Diagnostic analysis of the logistic model for pedestrian injury severity in traffic crashes. *Accident; Analysis and Prevention*, *39*(6), 1267–1278. doi:10.1016/j.aap.2007.03.017 PMID:17920851

Taniguchi, E., Thompson, R. G. & Yamada, T. (2013). Concepts and Visions for Urban Transport and Logistics Relating to Human Security. *Urban Transportation and Logistics: Health, Safety, and Security Concerns*, 1.

Tarawneh, M. S. (2001). Evaluation of pedestrian speed in Jordan with investigation of some contributing factors. *Journal of Safety Research*, *32*(2), 229–236. doi:10.1016/S0022-4375(01)00046-9

Tay, R., Choi, J., Kattan, L., & Khan, A. (2011). *A multinomial logit model of pedestrian–vehicle crash severity*. Academic Press.

Team, R. C. R. (2014). A language and environment for statistical computing. R Foundation for Statistical Computing.

Team, R. C. R. (2014a). A language and environment for statistical computing. R Foundation for Statistical Computing.

Team, R. C. R. (2014b). *A language and environment for statistical computing*. Vienna, Austria: R Foundation for Statistical Computing.

TfL. (2010). Measuring Public Transport Accessibility Levels. *Transport for London*. Available: http://www.webptals.org.uk/

Thurston, D. L., & Sun, R. (1994). Machine Learning User Preferences for Structural Design. *Computer-Aided Civil and Infrastructure Engineering*, *9*(3), 185–197. doi:10.1111/j.1467-8667.1994.tb00372.x

Toran Pour, A., Moridpour, S., Tay, R., & Rajabifard, A. (Eds.). (2016). A Partial Proportional Odds Model for Pedestrian Crashes at Mid-Blocks in Melbourne Metropolitan Area. In *MATEC Web of Conferences*. EDP Sciences. 10.1051/matecconf/20168102020

Toran Pour, A., Moridpour, S., Tay, R., & Rajabifard, A. (2017). Modelling pedestrian crash severity at mid-blocks. Transportmetrica A. *Transportation Science*, *13*(3), 273–297.

Transport, D. o. (2009). *Victorian Integrated Survey of Travel and Activity*. Available: http://www5.transport.vic.gov.au/

Transport, D. O. (2009). *Victorian Integrated Survey of Travel and Activity*. Available: http://www5.transport.vic.gov.au/

Tulu, G. S., Washington, S., Haque, M. M., & King, M. J. (2015). Investigation of pedestrian crashes on two-way two-lane rural roads in Ethiopia. *Accident; Analysis and Prevention*, *78*, 118–126. doi:10.1016/j.aap.2015.02.011 PMID:25770907

Vale, D. S. (2013). Does commuting time tolerance impede sustainable urban mobility? Analysing the impacts on commuting behaviour as a result of workplace relocation to a mixed-use centre in Lisbon. *Journal of Transport Geography*, *32*, 38–48. doi:10.1016/j.jtrangeo.2013.08.003

Van Dyck, D., Cardon, G., Deforche, B., Sallis, J. F., Owen, N., & De Bourdeaudhuij, I. (2010). Neighborhood SES and walkability are related to physical activity behavior in Belgian adults. *Preventive Medicine*, *50*(Supplement), S74–S9. doi:10.1016/j.ypmed.2009.07.027 PMID:19751757

Van Holle, V., Van Cauwenberg, J., Deforche, B., Goubert, L., Maes, L., Nasar, J., ... De Bourdeaudhuij, I. (2014). Environmental invitingness for transport-related cycling in middle-aged adults: A proof of concept study using photographs. *Transportation Research Part A, Policy and Practice*, *69*, 432–446. doi:10.1016/j.tra.2014.09.009

Verzosa, N., & Miles, R. (2016). Severity of road crashes involving pedestrians in Metro Manila, Philippines. *Accident; Analysis and Prevention*, *94*, 216–226. doi:10.1016/j.aap.2016.06.006 PMID:27340839

VicRoads. (2014). *Traffic monitor 2012-13*. Melbourne: Roads Corporation of Victoria.

VicRoads. (2015). *Principal Bicycle Network (PBN)*. Vicoria, Australia: Victorian Government Data Directory.

Wang, C.-H., & Chen, N. (2015). A GIS-based spatial statistical approach to modeling job accessibility by transportation mode: Case study of Columbus, Ohio. *Journal of Transport Geography*, *45*, 1–11. doi:10.1016/j.jtrangeo.2015.03.015

Wang, J., Zheng, Y., Li, X., Yu, C., Kodaka, K., & Li, K. (2015). Driving risk assessment using near-crash database through data mining of tree-based model. *Accident; Analysis and Prevention*, *84*, 54–64. doi:10.1016/j.aap.2015.07.007 PMID:26319604

Weber, J. (2006). Reflections on the future of accessibility. *Journal of Transport Geography*, *14*(5), 399–400. doi:10.1016/j.jtrangeo.2006.06.005

Wey, W.-M., & Chiu, Y.-H. (2013). Assessing the walkability of pedestrian environment under the transit-oriented development. *Habitat International*, *38*, 106–118. doi:10.1016/j.habitatint.2012.05.004

WHO. (2013). *Pedestrian safety: a road safety manual for decision-makers and practitioners*. World Health Organization.

Wier, M., Weintraub, J., Humphreys, E. H., Seto, E., & Bhatia, R. (2009). An area-level model of vehicle-pedestrian injury collisions with implications for land use and transportation planning. *Accident; Analysis and Prevention*, *41*(1), 137–145. doi:10.1016/j.aap.2008.10.001 PMID:19114148

Wilde, G. J. S. (1976). Social Interaction Patterns in Driver Behavior: An Introductory Review. *Human Factors*, *18*(5), 477–492. doi:10.1177/001872087601800506

Wilsnack, R. W., Wilsnack, S. C., Kristjanson, A. F., Vogeltanz-Holm, N. D., & Gmel, G. (2009). Gender and alcohol consumption: Patterns from the multinational GENACIS project. *Addiction (Abingdon, England)*, *104*(9), 1487–1500. doi:10.1111/j.1360-0443.2009.02696.x PMID:19686518

Winters, M., Brauer, M., Setton, E. M., & Teschke, K. (2010). Built environment influences on healthy transportation choices: Bicycling versus driving. *Journal of Urban Health*, *87*(6), 969–993. doi:10.100711524-010-9509-6 PMID:21174189

Wong, D. (2009). The modifiable areal unit problem (MAUP). The SAGE handbook of spatial analysis, 105-123.

Wu, B. M., & Hine, J. P. (2003). A PTAL approach to measuring changes in bus service accessibility. *Transport Policy*, *10*(4), 307–320. doi:10.1016/S0967-070X(03)00053-2

Xie, Z., & Yan, J. (2008). Kernel Density Estimation of traffic accidents in a network space. *Computers, Environment and Urban Systems*, *32*(5), 396–406. doi:10.1016/j.compenvurbsys.2008.05.001

Yingying, Z., Danya, Y., Qiu, T. Z., Lihui, P., & Yi, Z. (2012). Pedestrian Safety Analysis in Mixed Traffic Conditions Using Video Data. Intelligent Transportation Systems. *IEEE Transactions on.*, *13*(4), 1832–1844.

Yu, R., & Abdel-Aty, M. (2014). Analyzing crash injury severity for a mountainous freeway incorporating real-time traffic and weather data. *Safety Science*, *63*, 50–56. doi:10.1016/j.ssci.2013.10.012

Zajac, S. S., & Ivan, J. N. (2003). Factors influencing injury severity of motor vehicle–crossing pedestrian crashes in rural Connecticut. *Accident; Analysis and Prevention*, *35*(3), 369–379. doi:10.1016/S0001-4575(02)00013-1 PMID:12643954

Zegeer, C., Stewart, J., Huang, H., & Lagerwey, P. (2001). Safety effects of marked versus unmarked crosswalks at uncontrolled locations: Analysis of pedestrian crashes in 30 cities. *Transportation Research Record: Journal of the Transportation Research Board*, 56–68.

Zheng, Y., Chase, T., Elefteriadou, L., Schroeder, B., & Sisiopiku, V. P. (2015). Modeling vehicle–pedestrian interactions outside of crosswalks. *Simulation Modelling Practice and Theory*, *59*, 89–101. doi:10.1016/j.simpat.2015.08.005

Zhu, X., & Lee, C. (2008). Walkability and Safety Around Elementary Schools: Economic and Ethnic Disparities. *American Journal of Preventive Medicine*, *34*(4), 282–290. doi:10.1016/j.amepre.2008.01.024 PMID:18374241

Index

H

I

L

M

N

O

P

R

S

T

U

V

W

www.ingramcontent.com/pod-product-compliance
Lightning Source LLC
LaVergne TN
LVHW062124200926
848484LV00007B/1760